Editor: Burr Angle
Art Director: Lawrence Luser
Editorial Assistant: Marcia Stern
Art and Layout: Wells S. Marshall, III

KALMBACH BOOKS

First printing, 1985. Second printing, 1989. Third printing, 1993.

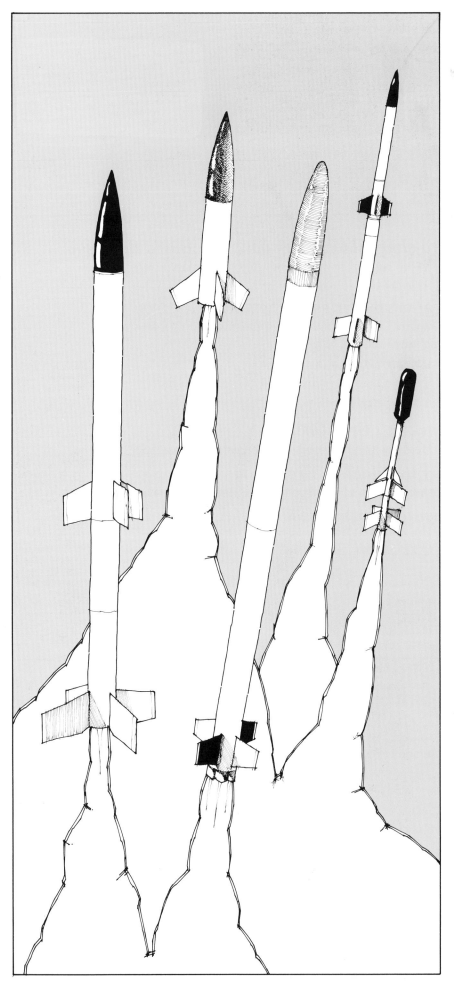

Introduction

By Michael A. Banks

Welcome to the "second stage" of model rocketry! If you've been in the hobby for a while, you may be finding yourself a little bored with watching your models go up and come down, however successfully. You've probably even scratchbuilt a few models of your own design, but may wonder what else the hobby has to offer. In this book, with the help of several important contributors, I want to show just how much there is waiting for the model rocket enthusiast. You'll see that the range of activities in advanced model rocketry is almost limitless.

For example, high-power rocketry offers the thrill of flying larger models to ultrahigh altitudes. Aerial photography offers a rewarding challenge, as well as the opportunity for serious research. Electronics applications also offer the opportunity for research, both in designing and assembling equipment and using it to gather information. There are many applications for computers in model rocketry. Scale modeling — for competition or just for personal satisfaction — offers its own rewards.

All of these activities add up to fun, which is what model rocketry is really all about.

Building materials

Materials for advanced model rockets are fairly easy to obtain. Body tubes, balsa and plywood fin stock and nose cones, plastic fin and nose cone sets, adapters, engine mounting tubes, parachutes and streamers, launch lugs, decals, and more are sold by Estes and several other firms, and were also manufactured by Centuri. (Although Centuri is no longer in business, I've included descriptions of several Centuri products throughout this book because these may still be available in some hobby shops and from specialist mail-order suppliers.)

You may, of course, wish to give a model a different appearance than can be provided with standard materials. If so, you will find that parts from wood, metal, and injection-molded or vacuum-formed plastic model kits — be they aircraft, spacecraft, ships, or cars — can be useful in customizing model rockets.

Model railroad accessories in several scales are another source of materials for building model rockets. The part you need to simulate an antenna strut on a scale model rocket may be mas-

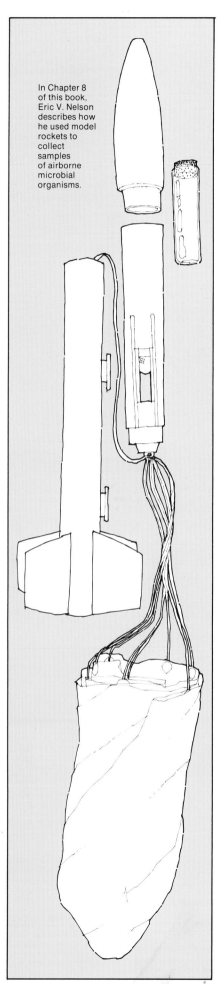

In Chapter 8 of this book, Eric V. Nelson describes how he used model rockets to collect samples of airborne microbial organisms.

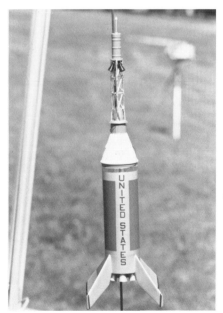

Skilled modelers research and scratchbuild scale versions of full-size prototypes; they sometimes even build scale models of launch towers and escape capsules.

querading as a piece of hardware on an N scale locomotive! Obtain a few catalogs from model railroad suppliers and plastic model manufacturers; you'll be pleasantly surprised at how much is available.

If all else fails, use your imagination. The endless variety of plastic packaging materials can supply quite a few model rocket components. The containers used to package L'eggs panty hose, for example, can easily be fashioned into nose cones or boattails. Plastic caps from mouthwash and other bottles can be pressed into service as nose cones or reducers. If you plan to do quite a bit of modeling — especially scale — it is a good idea to accumulate plastic lids, caps, bottles, and other decorative packaging materials. You never know when that odd-shaped bit of plastic you would normally throw out will be of use.

Craft shops are another good source of building materials. Most craft shops stock a fantastic variety of wood, metal, and Styrofoam shapes.

Usually overlooked, electronic components can serve as trim on any model rocket.

Despite the many sources of parts for the scavenger, you may find it necessary to build your own detail components. A vacuum-forming machine is useful, but may not be feasible for financial or other reasons. Keeping an ample supply of styrene sheets, rods, tubes, and other shapes on hand is a good alternative. With a steady hand, patience, and the proper heating and cutting tools, you can form styrene into almost any shape.

You can also scratchbuild major structural components such as body tubes and nose cones. Balsa nose cones can be turned on lathes or drills and body tubes can be made from rolled kraft paper tape.

Safety in advanced model rocketry

Safety in model rocketry cannot be overemphasized. The hobby has an enviable safety record (statistically better than baseball) and we want to keep it that way. Of course, we want to prevent injuries to anyone involving model rockets, but we also wish to forestall any attempt to outlaw the hobby.

Model rocketry has been defined by the Federal Aviation Administration in *Federal Aviation Regulations, Part 101*, and this helps impart some measure of legitimacy in the eyes of local authorities. The FAA definition states that a model rocket may weigh no more than 16 ounces, including not more than 4 ounces of propellant.

In addition, the National Fire Protection Association has established a model state fireworks law, in which model rockets are declared not to be fireworks.

It has taken quite a bit of effort by individuals such as G. Harry Stine and organizations such as the National Association of Rocketry to establish model rocketry as a safe, legal activity. A few cases of property and personal damage could change this.

With that in mind, re-read the Model Rocketry Safety Code one more time. This code is included with model rocket kits, but it bears repeating here.

In advanced model rocketry, you will frequently use experimental and untested technology, so extra attention to safety is a must. You should build safety into your models from the start. This means thinking out all possible effects of the project you are developing, as well as determining just what sorts of malfunctions could occur. Once you have done this, assume that these things will go wrong and prepare for each eventuality.

Select your building materials, and design and build your high-power models with strength in mind, but stay within weight limits.

Test each model before flight. This means checking for adequate stability by balancing the rocket at the end of a piece of string and swinging it in a circle or by using any of the methods for determining a model's center of gravity and center of pressure.

Handle all engines, but especially high-power engines, carefully. Don't drop them, and don't subject them to high temperatures (such as may be present in the trunk of a car on a hot day). Use only engines approved by the National Association of Rocketry — don't try to make your own.

Conduct test flights far from houses and other property and with as few spectators as possible. Ideally, only the rocketeer conducting the experiments and a range crew and recovery crew (sometimes these are the same) should be present. Conversely, never fly untested designs alone.

In addition to conducting tests as far from people and property as possible, all of the standard range safety practices should be observed. Here's a typical preflight checklist:

☐ 1. Be sure that the flying field is clear of ground obstacles and that no low-level obstacles such as power lines are present.

☐ 2. Inspect the launch equipment (you should be using a launch controller with a safety interlock).

☐ 3. Inspect the rocket. Are the fins and nose cone in place and is the recovery system prepped? Is the engine properly in place?

☐ 4. Check the wind direction. You will, of course, have a good idea of what the weather will be like before you fly, but be alert to the possibility of sudden gusts of wind. Adjust the angle of your launcher to compensate for any breezes in the launch area.

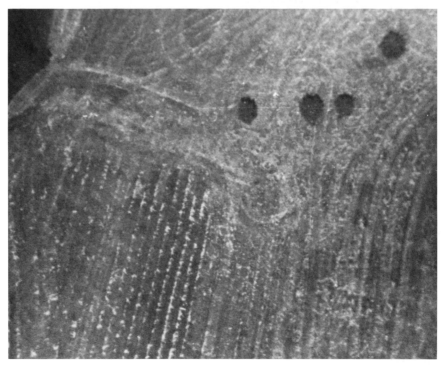

You'll also discover how to use the Estes Astrocam 110 to take color photos at altitudes ranging from several hundred to several thousand feet.

5

Keep in mind that high-performance model rockets require a large flying site clear of houses or other buildings. If possible, the site should also be free or nearly free of vegetation.

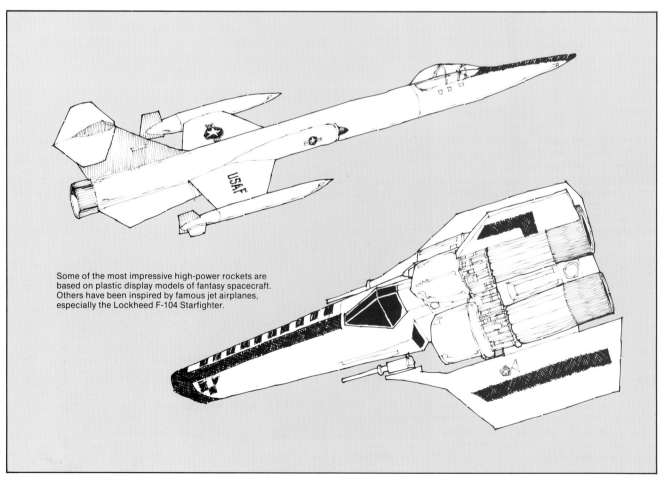

Some of the most impressive high-power rockets are based on plastic display models of fantasy spacecraft. Others have been inspired by famous jet airplanes, especially the Lockheed F-104 Starfighter.

Photos, patches, and scale models can form the basis for a striking display demonstrating man's achievements in space.

☐ 5. Scan the sky for low-flying aircraft. Even the smallest model rocket can be dangerous to aircraft.

☐ 6. Look over the launch range to make sure that no people have wandered into the recovery area.

☐ 7. Make certain that your recovery crew and observers are in place.

☐ 8. Check once more for low-flying aircraft.

☐ 9. Perform a 10-second countdown, counting backwards from 10 and launching at 0. While you are calling the count, keep an eye out for any changes in wind, the presence of aircraft, or any other factors which might require aborting the launch.

And don't hesitate to abort! It is better to wait a few minutes than to risk hurting someone or damaging something. All sorts of strange things can happen just before a launch. I've seen the following: A launch pad fell over for no reason at all; a small breeze built up to a 30-mph gust in a few seconds; a dog wandered into the launch area and upset the launcher; a spectator ran up to the launcher for a better view; a low-flying aircraft appeared silently over the horizon.

Remember safety is your watchword. Be alert, observe safety practices, and use common sense.

MODEL ROCKETRY SAFETY CODE

1. **Construction.** My model rockets will be made of lightweight materials such as paper, wood, plastic, and rubber, without any metal as structural parts.

2. **Engines.** I will use only pre-loaded factory-made NAR safety-certified model rocket engines in the manner recommended by the manufacturer. I will not change in any way nor attempt to reload these engines.

3. **Recovery.** I will always use a recovery system in my model rockets that will return them safely to the ground so that they may be flown again.

4. **Weight limits.** My model rocket will weigh no more than 453 grams (16 ounces) at liftoff and the engines will contain no more than 113 grams (4 ounces) of propellant.

5. **Stability.** I will check the stability of my model rockets before their first flight, except when launching models of already proven stability.

6. **Launching system.** The system I use to launch my model rockets must be remotely controlled and electrically operated, and will contain a switch that will return to "off" when released. I will remain at least 10 feet away from any rocket that is being launched.

7. **Launch safety.** I will not let anyone approach a model rocket on a launcher until I have made sure that either the safety interlock key has been removed or the battery has been disconnected from my launcher.

8. **Flying conditions.** I will not launch my model rocket in high winds, near buildings, power lines, tall trees, low-flying aircraft, or under any conditions which might be dangerous to people or property.

9. **Launch area.** My model rockets will always be launched from a cleared area free of any easy-to-burn materials, and I will only use nonflammable recovery wadding in my rockets.

10. **Jet deflector.** My launcher will have a jet deflector device to prevent the engine exhaust from hitting the ground directly.

11. **Launch rod.** To prevent accidental eye injury I will always place the launcher so that the end of the rod is above eye level or cap the end of the rod with my hand when approaching it. I will never place my head or body over the launching rod. When my launcher is not in use I will always store it so that the launch rod is not in an upright position.

12. **Power lines.** I will never attempt to recover my rocket from a power line or other dangerous place.

13. **Launch targets and angles.** I will not launch rockets so their flight path will carry them against targets on the ground, and will never use an explosive warhead nor a payload that is intended to be flammable. My launching device will always be pointed within 30 degrees of vertical.

14. **Pre-launch test.** When conducting research activities with unproven designs or methods, I will, when possible, determine their reliability through pre-launch tests. I will conduct launchings of unproven designs in complete isolation from persons not participating in the actual launching.

Most high-power rocketry supplies are produced in limited quantities by individuals and small firms; they are usually sold by mail order or at swap meets.

1. High-power model rockets

By Michael A. Banks

I find high-power rockets extremely useful in aerial photography because I can carry an Astrocam to a much higher altitude than with less powerful engines. Other rocketeers are interested in attaining maximum altitudes or speeds or in lofting heavier than normal research payloads. Still others find that high-power engines are ideal for flying large models which, although they are within legal weight limits, require extra power because their shape causes lots of drag.

A high-powered model rocket is propelled by one or more engines that provide a total impulse of more than 20 newton-seconds.

Thus, the use of E and F engines (which are 40 and 80 newton-seconds, respectively) puts you into the area of high-power rocketry. Clustering D engines or using any combination of engines (such as three C engines) firing simultaneously also constitutes high-power model rocketry.

Manufacturers occasionally offer G or even H engines but these are illegal in some localities and require a special permit in most others. I suggest you stay away from them.

I've listed several firms that sell E and F engines on page 63, but be aware that small businesses offering model rocket supplies come and go rapidly. Your best sources of information about the availability of specialized products are magazines such as *American Spacemodeling* (formerly *The Model Rocketeer*) and club newsletters.

Restrictions

The weight limit of 16 ounces (453 grams) for a model rocket restricts just how many engines you can use in a model rocket and thus the total impulse attainable. Obviously, clustering ten D engines would give you a high-powered rocket, but the engines themselves would be over the limit of 453 grams for a model rocket! Clustering that many engines would also be dangerous because one or more of the engines would probably fail to ignite and this could make for an unstable rocket.

Further, as I mentioned in the Introduction, the FAA states that a model rocket may have no more than 4 ounces (113 grams) of propellant. Thus, a cluster of more than four D engines would be illegal or require special permits in most areas. (Besides, clustering D engines is very inefficient.)

As far as individual engines are concerned, the National Association of Rocketry (NAR) and the National Fire Protection Association have declared that a single model rocket engine may not have more than 80 newton-seconds of total impulse and may not contain more than 62.5 grams of propellant.

What kind of practical limitations do we have for total impulse then? Well, consider a hypothetical clustering of F engines. You might cluster two F engines for a total impulse of 160 newton-seconds and, since a typical F engine has a propellant weight of 42 grams, stay within the weight limit of 113 grams of total propellant weight for the rocket. If, however, you added a third engine, you would be exceeding the weight limit for propellant (3 x 42 = 126 grams total propellant weight).

Composite engines

Composite engines have a propellant made from a blend of ammonium perchlorate (AP) and other materials; their specific impulse is about 2.5 times greater than conventional black powder engines of equal weight.

For example, a (hypothetical) black-powder engine containing the maximum permissible 62.5 grams of propellant would achieve a total impulse of slightly over 50 newton-seconds. The case would weigh nearly as much as the propellant, say 43 grams, for a gross weight of 105.5 grams.

Now, let's compare this with what can be done with composite propellant. A typical composite E engine achieves a total impulse of 40 newton-seconds with a propellant weight of 21 grams. This is a little over one-third the weight of the amount of black powder needed to produce 50 newton-seconds of thrust. Further, the composite E engine case weighs only 30 grams. So in a package weighing only 51 grams, we have far greater efficiency and nearly as much thrust as we'd obtain from a black powder engine at 105.5 grams.

Composite engine design

Composite propellant consists primarily of ammonium perchlorate, which functions as the oxidizer. A binder which also serves as the fuel element is also present, as are other chemicals that determine the burning rate. The cases are lightweight fiberglass or phenolic tubing.

The propellant in composite model rocket engines is similar in many respects to that used in the Solid Rocket Boosters (SRBs) in the Space Shuttle.

It is theoretically possible to attain chamber pressures of 1,000 pounds per square inch absolute (psia) with AP composite similar to that used in model rocket engines but the heavy case required to contain such pressure would take the engine weight far beyond the 113 gram limit. Therefore, composite engine manufacturers keep the chamber pressure down to 300 to 400 psia, which is still three to four times that of black powder engines.

Physically, a composite model rocket engine, Fig. 1-1, is similar to a black powder engine, with the exception of the ejection charge, which does not contact the propellant.

Table 1-1 presents comparative data

These are Estes black powder engines in cardboard cases; top to bottom, a mini-engine, a regular engine, and a D engine. The D engine is 2.75″ long, .690″ in diameter. Aero Tech and possibly other firms offer high-power composite propellant E engines in D-sized cases so that you can often convert from D to E power simply by changing engines.

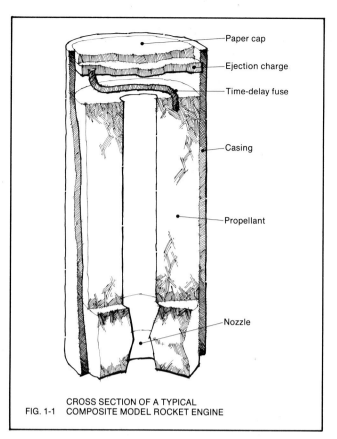

CROSS SECTION OF A TYPICAL
FIG. 1-1   COMPOSITE MODEL ROCKET ENGINE

on black powder and composite engines.

High-strength materials and special building techniques

I can't overstress the need for great strength in high-power model rockets, where you're dealing with total and average impulse levels that are two to eight times higher than with smaller engines.

Structural integrity in any aerospace vehicle begins with high-strength materials. This is true of the Space Shuttle, the latest jet fighter aircraft, and model rockets. In model rocketry, as with full-scale aerospace vehicle design and construction, we are also concerned with lightweight materials because any unnecessary weight can degrade performance. We are, perhaps, even more concerned with weight in our model rockets than full-size rocket designers are, since they can often rather easily add power and are not limited by absolute maximum weight restrictions.

We have to make trade-offs in designing and building model rockets, just as full-size rocket engineers do. A trade-off, of course, is accepting lesser efficiency in one area for greater efficiency in another in an attempt to achieve overall optimum performance levels. For example, an engineer working on a rocket that will deliver a certain payload into earth orbit may have to trade payload capacity for structural strength if the rocket is to withstand the stresses of launch.

Conventional body tubes, you will find, can usually take the stresses of high-power launches quite well, but when you fly rather long high-power rockets, you may find that body tubes crimp during liftoff. The solution is to use body tubes with thicker walls; instead of, for example Estes BT-70, which has a wall thickness of .021″, you may want to use body tubes provided by some high-power rocket manufacturers which have a wall thickness of perhaps .05″. The extra thickness will add weight to your model, but you are flying with higher than normal impulse levels and the strength provided by the extra thickness will ensure a successful flight.

The materials you use for fins and engine thrust rings will also require a trade-off of weight for strength. Balsa parts often cannot withstand the stress of a high-power engine and may shred during launch. So, instead of using balsa for fins and engine thrust rings, you should use basswood, which is not quite as light as balsa, but stronger, or, better still, plywood.

The type of plywood I'm talking about here isn't what you find at lumberyards. I'm referring to aircraft plywood, the kind used by model aviators and that is sold by most hobby shops. This plywood, usually made of birch, comes in thicknesses from 1/16″ through 1/4″. It is very, very strong, but heavier than basswood.

That good old white glue you use when assembling conventional kits just won't do in building high-power rockets. Components such as fins will be

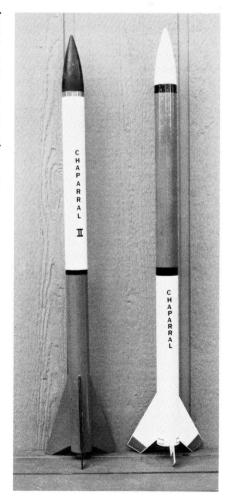

Rich Cardillo's Chaparral and Chaparral II are sport birds for E and F engines; plans are on pages 14 and 15.

TABLE 1-1 WEIGHT AND PERFORMANCE FACTORS OF HIGH-POWER BLACK POWDER AND COMPOSITE ENGINES

ENGINE	TOTAL IMPULSE	AVERAGE IMPULSE	THRUST DURATION	TOTAL WEIGHT	PROPELLANT WEIGHT
D12 Black powder	17 N-Seconds	12.05 N-Seconds	1.41 Seconds	42 Grams	24.9 Grams
E20 Composite	40 N-Seconds	20 N-Seconds	2 Seconds	52 Grams	21 Grams
F40 Composite	80 N-Seconds	40 N-Seconds	2 Seconds	78 Grams	42 Grams

The exhaust from high-power engines will quickly eat through the thin steel used on ordinary blast deflectors, so keep a good supply of spares on hand or make a sturdier deflector from ³⁄₁₆″ or thicker steel plate.

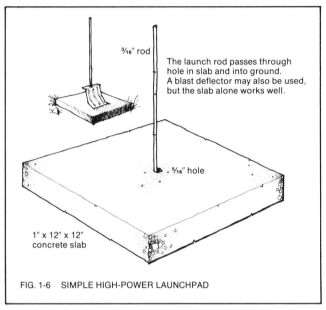

The launch rod passes through hole in slab and into ground. A blast deflector may also be used, but the slab alone works well.

³⁄₁₆″ rod

⁵⁄₁₆″ hole

1″ x 12″ x 12″ concrete slab

FIG. 1-6   SIMPLE HIGH-POWER LAUNCHPAD

heavier, of course, and any part that you glue in place will be subjected to much higher stresses during flight than with conventional rockets.

Glues work by adhering strongly to two different surfaces, but the bond itself (that is, the glue between the objects) must also be strong. In the case of white glue, yellow glue, or glues based on animal protein extracts these bonds are sometimes weak or brittle.

Epoxies, on the other hand, are much stronger, both in adhering to surfaces and in their bonds. In addition to being strong and tough, epoxies resist corrosion, withstand temperatures up to several hundred degrees Fahrenheit, and adhere well to most metals, wood, paper, and some plastics — to almost all of the materials we use in model rockets.

With all of these qualities, epoxies are the best type of glue for high-power model rocketry.

There are two kinds of epoxy glue: regular and quick-setting. The regular type sets up (becomes tacky and capable of holding objects together if there is no stress on the parts) in about 30 minutes. The quick-setting variety sets up in 3 to 5 minutes. Both types achieve full strength (maximum hardness and bonding strength) in 24 hours.

The regular epoxy is runny and doesn't "grab" components such as fins the way thicker glues do, so you may have to devise jigs to hold fins until the epoxy thickens as it sets up. An alternative is to tack glue the fins with quick-setting epoxy or white glue and later reinforce the joint with regular epoxy.

I haven't used any of the relatively new cyanoacrylate instant glues extensively in assembling models, but have found them to be excellent for field repairs such as fixing a cracked fin.

If you are working with styrene or acrylic plastic parts, be aware that there are two major types of plastic glues. Solvent-type plastic glue (also known as "liquid plastic cement") is a clear liquid that fuses plastic pieces together with a solvent action. When the solvent evaporates and the plastic hardens, you are left with two pieces of plastic that are actually welded together. I don't like working with liquid plastic cement because applying the proper amount is tricky. Too much cement and you are left with plastic parts that have dissolved or distorted into unusable shapes. Too little and there is no bonding or very weak bonding.

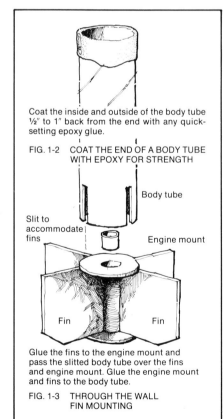

Coat the inside and outside of the body tube ½″ to 1″ back from the end with any quick-setting epoxy glue.

FIG. 1-2   COAT THE END OF A BODY TUBE WITH EPOXY FOR STRENGTH

Body tube

Slit to accommodate fins

Engine mount

Fin   Fin

Glue the fins to the engine mount and pass the slitted body tube over the fins and engine mount. Glue the engine mount and fins to the body tube.

FIG. 1-3   THROUGH THE WALL FIN MOUNTING

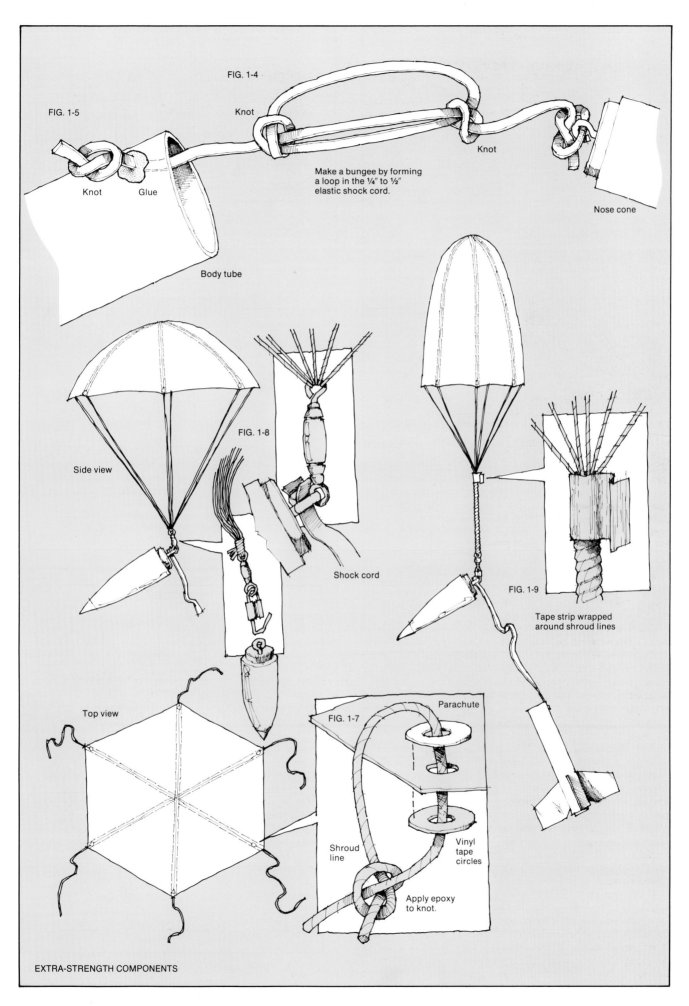

FIG. 1-4

FIG. 1-5

Knot

Knot

Knot

Knot   Glue

Make a bungee by forming
a loop in the ¼" to ½"
elastic shock cord.

Nose cone

Body tube

Side view

FIG. 1-8

Shock cord

FIG. 1-9

Tape strip wrapped
around shroud lines

Top view

FIG. 1-7

Parachute

Shroud
line

Vinyl
tape
circles

Apply epoxy
to knot.

EXTRA-STRENGTH COMPONENTS

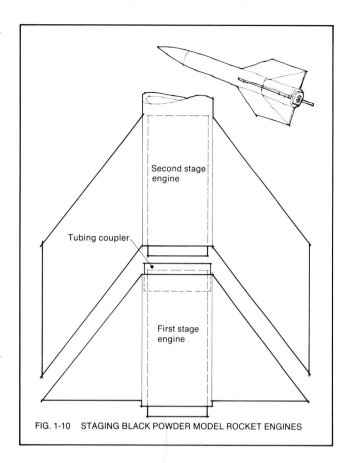

FIG. 1-10   STAGING BLACK POWDER MODEL ROCKET ENGINES

Second stage engine

Tubing coupler

First stage engine

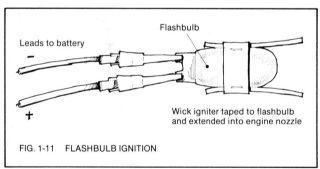

Flashbulb

Leads to battery

Wick igniter taped to flashbulb and extended into engine nozzle

FIG. 1-11   FLASHBULB IGNITION

Capacitor

Leads to igniter

Leads to battery

Mercury switch

Use a 1000-microfarad capacitor with a voltage rating equal to or greater than that of the battery used in your launcher. Install the mercury switch so that the bottom (the location of the leads) is "up" with reference to the rocket's direction of travel. At apogee, as the rocket begins to decelerate, the mercury in the switch will be thrown upward, completing the circuit and allowing the current from the capacitor to discharge into the igniter.

The capacitor must be charged before launching.

FIG. 1-12   SECOND STAGE IGNITION SYSTEM

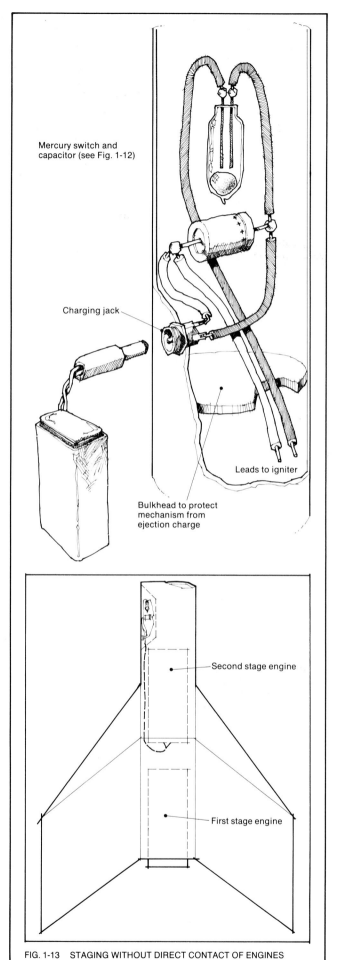

Mercury switch and capacitor (see Fig. 1-12)

Charging jack

Leads to igniter

Bulkhead to protect mechanism from ejection charge

Second stage engine

First stage engine

FIG. 1-13   STAGING WITHOUT DIRECT CONTACT OF ENGINES

The other type of plastic glue is tube-type model cement (also known as "tube glue"). I like to use tube glue because it sets up slowly, allowing me to move plastic parts as I wish, and it makes a strong bond.

You may wish to coat portions of the body tube with epoxy for added strength. For example, I always coat the top of the body tube where the nose cone fits with epoxy inside and out for about an inch back, Fig. 1-2. This adds strength to an area that undergoes a lot of stress when the nose cone is ejected and sometimes upon landing as well.

You may also coat the entire body tube with epoxy. This strengthens the body tube and, if you sand the surface very smooth, reduces drag significantly. Unfortunately, the epoxy coating also adds weight to the rocket, so you are again looking at a trade-off — in this case, extra weight for increased strength and reduced drag.

Nose cones can be coated with epoxy or with balsa fillercoat. If the nose cone is made of plastic or hardwood, you probably won't need a coating, unless you want to use a dark or light undercoating as a primer for the color coats of paint.

Fins can also be treated with balsa fillercoat or epoxy. I like to paint fins with fillercoat (except when I'm using plywood fins, which are already pretty smooth and well sealed) and then reinforce the leading and trailing edges with a little epoxy, since these edges receive the most abuse during launch and recovery.

Remember that all of these coatings are going to add weight. You can reduce the additional weight by careful sanding, beginning with medium-grit sandpaper and progressing to increasingly fine grits. You should be sanding all coated and uncoated surfaces smooth anyway to help in painting and to reduce drag.

One of the major techniques for building in strength in high-power rocket construction is the through-the-wall fin-mounting technique perfected by Rich Cardillo. The fins extend through the wall of the body tube and are epoxied directly to the engine mounting tube. The fins and engine mount are assembled as one unit, Fig. 1-3, and then put into the body tube, using slits cut in the body to accommodate the fins. Epoxy is placed on the mounting rings surrounding the engine mounting tube before the assembly is put into the body tube. After everything is in place and the first application of glue has dried, additional epoxy can be placed around the inside and outside of the rear mounting rings.

Once again, note that the rings should be made of basswood or plywood, not balsa. You don't want the engine mount to take off through the center of the rocket!

The points where the fins pass through the body tube should also be epoxied; build up fillets of glue to reduce drag. Slits below the fins can be sealed with body tube material glued on from inside.

Shock cords should also be extra strong, so use ¼" or ½" elastic, which is sold by fabric stores. You can further enhance the strength of shock cords by looping the elastic into a bungee, Fig. 1-4. Shock cords should be anchored with epoxy and may be extended through the body tube for extra strength, Fig. 1-5.

Launch rods

Since most high-power rockets are longer than conventional model rockets, you will need a longer launch rod. The added weight of these rockets also dictates a larger diameter rod. The minimum launch rod acceptable for use with high-power rockets is the ³⁄₁₆"-diameter, 36"-long Estes Maxi-rod. Anything smaller is liable to bend, sending the rocket on a poorly angled flight path.

For rockets that are 4 feet or longer, it's best to use a ¼"- or even ⅜"-diameter steel or aluminum rod 4 to 6 feet long. Anchor the rod with a concrete block or use a launchpad like that in Fig. 1-6. Note that conventional blast deflectors won't stand up to repeated use with D, E, or F engines. Have an ample supply of spare deflectors on hand or make your own from ³⁄₁₆" sheet steel.

Large launch rods won't accommodate the conventional size launch lug, which has an inside diameter of ⁵⁄₃₂". Estes sells ³⁄₁₆"-diameter launch lugs that fit the Maxi-rod.

For larger rods, you will have to find suitably sized cardboard or Mylar plastic cylinders or you can clean out used mini-engine cases and install these as lugs.

Choosing a launch site

Obviously, high-power rockets require a larger recovery area than conventional model rockets — the higher a rocket flies, the farther it is likely to drift during descent. A good rule of thumb for selecting a field is to make certain that its shortest dimension is equal to half of the maximum altitude to which you are going to fly. Thus, if you expect to fly high-power rockets to altitudes of 2,000 feet, a field with a minimum dimension of 1,000 feet will be adequate for recovery.

A high-power rocket field should be located as far as possible from subdivisions or other densely populated areas. This will minimize the number of onlookers you'll attract as well as give your rockets a relatively safe area to land in should they drift out of the recovery area. (Irate homeowners sometimes keep any rockets they find in their backyards and may even call the police.)

Recovery techniques

There are two common recovery techniques for high-power rockets: parachute and streamer. Glide recovery is possible but exceedingly difficult. I have seen some E-powered boost-gliders fly successfully but they need a square mile or more of recovery area. Tumble and featherweight recovery are out of the question; high-power model rockets are much too large and heavy for these systems to work. A typical high-power model free falling from 1,000 feet would be damaged beyond repair and do considerable damage to anything it hit on impact.

Figure 1-7 illustrates my favorite

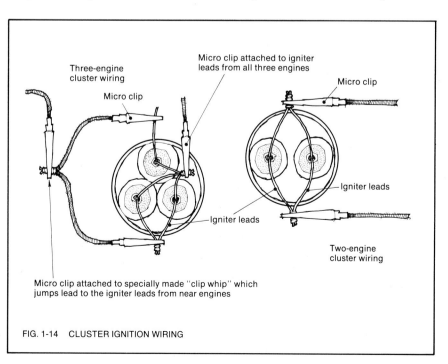

FIG. 1-14 CLUSTER IGNITION WIRING

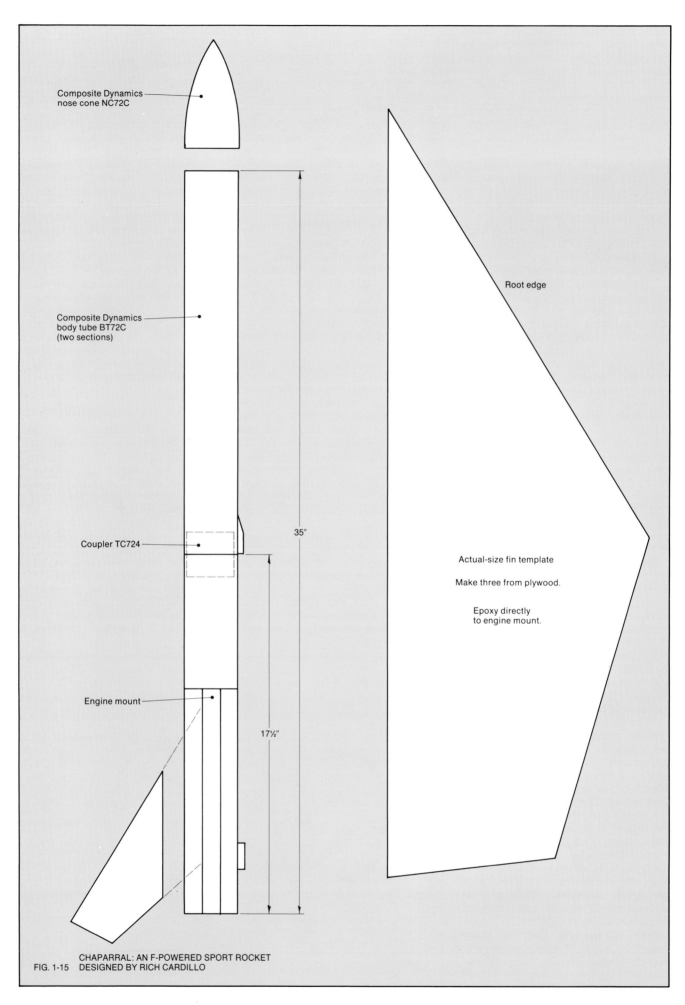

Composite Dynamics
nose cone NC72C

Composite Dynamics
body tube BT72C
(two sections)

Coupler TC724

Engine mount

35″

17½″

Root edge

Actual-size fin template

Make three from plywood.

Epoxy directly
to engine mount.

CHAPARRAL: AN F-POWERED SPORT ROCKET
FIG. 1-15   DESIGNED BY RICH CARDILLO

14

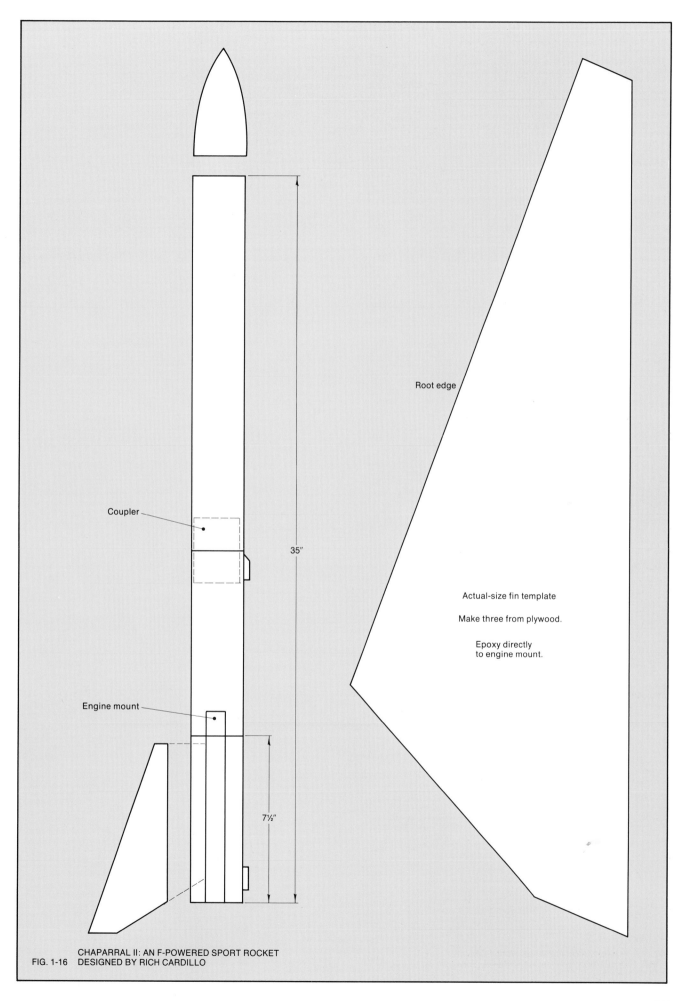

Root edge

Coupler

35″

Actual-size fin template

Make three from plywood.

Epoxy directly
to engine mount.

Engine mount

7½″

CHAPARRAL II: AN F-POWERED SPORT ROCKET
FIG. 1-16   DESIGNED BY RICH CARDILLO

15

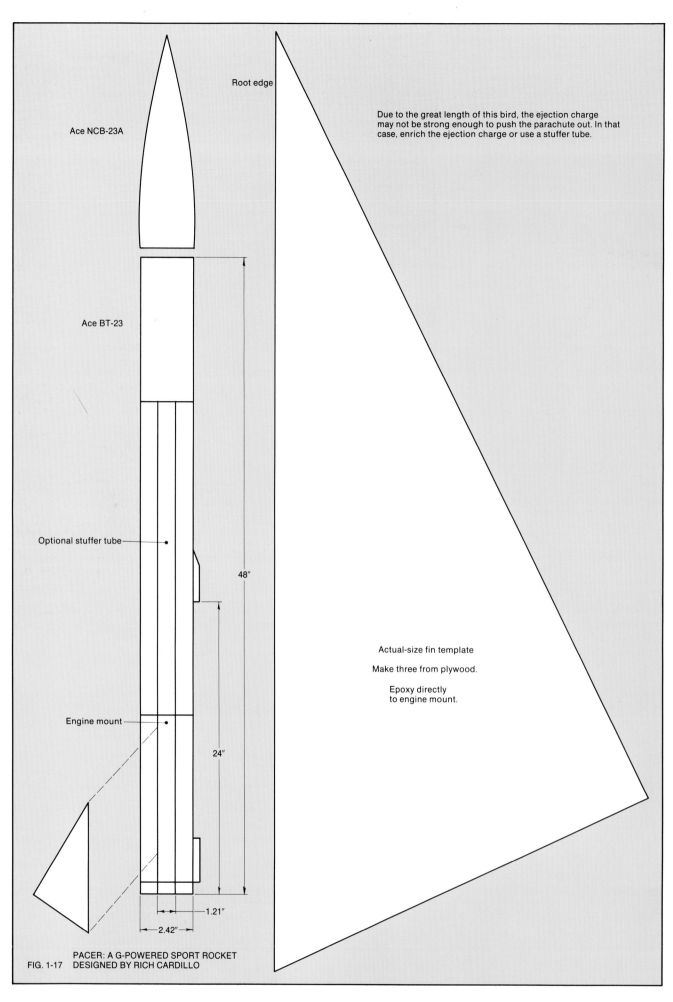

Root edge

Ace NCB-23A

Due to the great length of this bird, the ejection charge
may not be strong enough to push the parachute out. In that
case, enrich the ejection charge or use a stuffer tube.

Ace BT-23

Optional stuffer tube

48″

Actual-size fin template

Make three from plywood.

Epoxy directly
to engine mount.

Engine mount

24″

1.21″

2.42″

PACER: A G-POWERED SPORT ROCKET
FIG. 1-17   DESIGNED BY RICH CARDILLO

technique for attaching shroud lines to parachutes. You may use any type of line — heavy cotton string, monofilament fishing line, nylon cord — all work well. You can make parachutes from strong nylon (Ripstop nylon, used on full-size chutes, is excellent) or 3-mil or thicker plastic from heavy-duty trash bags.

The best parachutes are made completely from nylon with true canopies and sewn-in shroud lines that extend completely across the canopy, Fig. 1-8. They're extra strong and will provide a safe descent for just about any rocket.

Keep in mind that the higher the rocket goes the more it will drift. Size your parachutes accordingly, selecting smaller parachutes for rockets that will attain higher altitudes.

If you have a very large field, you may not be concerned about your rocket drifting during descent. In that case, you should go with a large-diameter parachute (perhaps 3 feet in diameter). However, if you plan to fly rockets to 2,000 feet or more, or your field is small, you will want to use smaller chutes or reef larger chutes, Fig. 1-9, for a more rapid descent.

In any case, attach the chute to the rocket with a snap swivel. Snap swivels are available in any store carrying fishing tackle, as they are often used to attach sinkers and leaders to fishing line.

Staging and clustering

Black powder D engines are staged by butting the top of the lower-stage engine against the bottom of the upper-stage engine. When the ejection charge of the first-stage engine ignites, the hot gases and particles thrown out enter the nozzle of the upper-stage engine, igniting it.

You should, of course, use a zero-delay booster engine for the lower stage.

Figure 1-10 will give you an idea of how to build a rocket to accommodate staging by butting the engines together, as will examining commercial kits for staged rockets that use this method. Note that the fins of the first and second stages fit together into one unit when the rocket is assembled for flight.

Staging by butting engines together will not work with composite engines. The ejection charge of a composite engine will not ignite another composite engine — you must use either igniters supplied by the engine manufacturer or flashbulb ignition, Fig. 1-11.

To stage composite engines, then, you must include a device which will set off the igniter for the second stage engine at the instant the first stage engine burns out. This can be accomplished in several ways. A sensor, for example, might be employed to react to the ejection charge, send a signal to a relay, and the relay in turn would complete the circuit between a power supply and the igniter.

Or you might use a timer set for the number of seconds between first stage engine ignition and burnout to complete the circuit between the power supply and the igniter. Either of these methods is expensive in terms of money (small electronic devices such as timing chips and miniature relays are not cheap) and weight. Too, you will need more than a little knowledge of electronics to design and build such devices.

Figure 1-12 presents an alternative made from a capacitor and mercury switch, both of which are simple and inexpensive. This device can also be used with black powder engines, particularly with models such as that in Fig. 1-13, in which the design of the rocket prohibits butting engines together, a frequent problem with scale designs.

The capacitor is charged before lift-off. The leads from the capacitor can be connected to a battery, then kept separated from one another as they are pushed back into the body of the rocket. It's best, though, to connect the capacitor leads to a jack mounted on the body tube. Insert a plug on the battery leads into the jack to charge the capacitor. The plug and jack approach looks more professional and eliminates the possibility of accidentally discharging the capacitor.

Clustering in high-power rocketry should only be done with black powder engines. Figure 1-14 shows how the leads to the igniters should be connected in parallel so that equal current is distributed to the igniters at the same time. This will ensure that the igniters fire simultaneously, igniting all engines.

Even one unignited engine in a cluster can make for a dangerous flight, so be sure that all connections are good and that the battery is fully charged when you attempt to launch a clustered rocket. NEVER CLUSTER COMPOSITE ENGINES.

Chaparral, Chaparral II, and Pacer

Here are plans, Figs. 1-15, 1-16, and 1-17, for three high-power rockets, all proven performers designed by Rich Cardillo. Both Chaparrals are for F engines; the Pacer is for G engines, and I recommend that you build and fly it only if you already have a lot of experience with E and F models and then only if you're certain G engines are legal in your area.

In each rocket epoxy the fins (make them from birch plywood) directly to the engine mount.

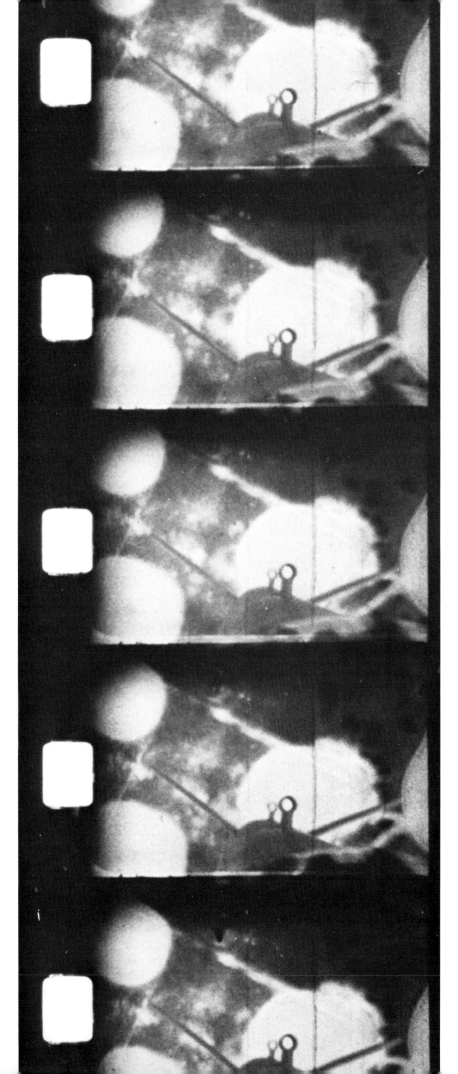

2. Aerial photography

By Michael A. Banks

One of the more obvious "serious" applications of model rocketry is aerial photography. After all, with a model rocket we have the capability for lifting a payload to a high altitude, so why not have an active payload, a payload that does something? These must have been the thoughts of early model rocketeers back in the late 1950s when the hobby was first getting started.

It was not until 1961 that a model rocketeer successfully flew a camera as a model rocket payload. Lewis Dewart, a Pennsylvania rocketeer, lofted a tiny Japanese camera strapped to the side of a model rocket that year. Upon ejection, the nose cone pulled a string attached to the shutter to shoot the picture. This was a chancy proposition because there was no guarantee that the lens would be pointed at the ground when the shutter snapped, but it worked.

Camroc

In 1965 Estes Industries introduced the first commercial model rocket camera payload. Dubbed the Camroc, it was specifically designed for use with a model rocket, and did its job well.

The Camroc was essentially a cylinder, 1.5″ in diameter, which was mounted atop a model rocket in place of the nose cone. Total weight was .5 ounce and the camera took one exposure per load of film.

The Camroc used specially packed circular pieces of Tri-X black-and-white film, ASA 400, which had to be pushed to ASA 1200 for development. Lens speed was f 11 and the focal length was 3″ (76 mm). The shutter speed was 1/1600 second. The fast shutter speed was necessary, of course, because the Camroc was moving rapidly when the shutter was released.

After processing by Estes, Camroc negatives yielded a 3″-diameter, somewhat grainy print.

Estes provided quite a bit of support over the years for Camroc users, including photo contests and information on photo interpretation and shooting stereopairs.

The Estes Cineroc carried 10 feet of Super 8 color film, which was sufficient for about 40 seconds of projection time. These five frames, enlarged many times, were taken from a rapidly ascending model rocket soon after launch — note the lugs and tail fins. All of the aerial photos in this chapter are black-and-white copies of color originals.

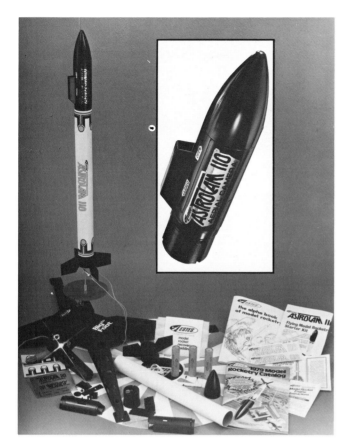

The current-model Estes Astrocam 110 is available separately or as part of a starter outfit that includes the kit for a Delta II launch vehicle, a launcher, and several engines.

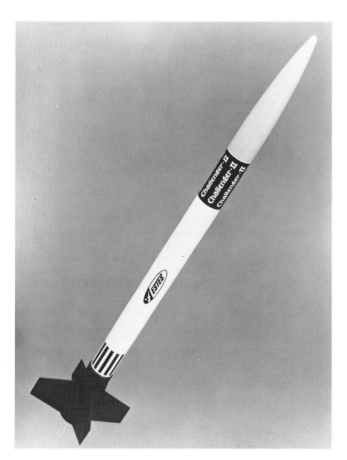

An Astrocam mounted in a D-powered Estes Challenger II can take pictures from over 1,200 feet. The camera uses ordinary 110 ASA 400 color film cartridges.

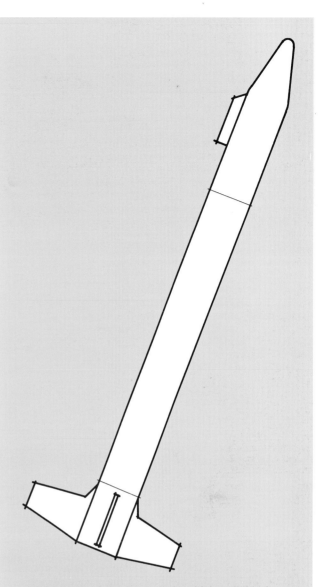

A high-power Astrocam launch vehicle suitable for use with 2.4 cm E and F engines such as Competition Dynamics E-20 and F-40 can be built from these Centuri parts:
- Series 13 body tube – No. 6010, ST1318
- Plastic fin unit – No. 5467, F413L
- D engine mount – No. 6056, EM913

Cut the body tube to 10″ (any excess length moves the center of gravity forward, due to the extra weight of the loaded Astrocam). Assemble the engine mount, leaving the engine clip out of the assembly.

After the glue on the engine mount parts has dried, install the engine mount in the body tube, leaving approximately ½″ of the mount extending from the tube. (Deleting the engine clip and extending the mount out of the body tube are necessary because the E and F engine casings are longer than the D engines for which the mount was designed.)

Next, cover an area of 1¾″ from the body tube with tube-type plastic cement (the bottom is the end in which you installed the engine mount). Carefully slide the plastic fin unit down from the top of the body tube all the way to the bottom of the tube. Wipe off excess glue and allow to dry.

After the glue has dried, install the recovery system. It's best to use a streamer instead of a parachute to limit drifting during recovery. A streamer 6″ wide and 10′ long is optimum.

When flying this rocket, use a ⅜″ launch rod, since higher-power engines can cause problems with a thin rod.

Obviously, a long delay time is necessary because the rocket will coast upward for much longer than a C- or D-powered booster of the same size.

Fig. 2-2 HIGH-POWER ASTROCAM LAUNCH VEHICLE

Aiming the Camroc was accomplished by using an engine with a delay time longer than normal for a rocket the size of the Camroc carrier. The excess delay time allowed the Camroc and its carrier to coast to apogee and then begin to fall, pointing nose down. By this time, the delay charge of the engine had burned through, causing ejection. At ejection, a string held in place by friction between the base of the Camroc and the forward end of the carrier's body tube was released. The string was attached to the shutter, and the release of the string caused a rubber band to pull the shutter over the lens.

Using an engine with a shorter delay time would sometimes result in a horizon photo, but a shot of blue sky was just as likely.

Control over the aiming of the Camroc was limited. Beyond experimenting with engine delay times the rocketeer could use a one- or two-stage carrier, or vary the launch angle to affect the flight path of the Camroc.

Estes offered the Camroc, along with one- and two-stage boosters, until 1976.

Cineroc

Aerial motion photography was the next step beyond still photos. In 1962, two New York rocketeers flew the first model rocket-borne movie camera. Using a spring-wound 8 mm Bolsey B-8, they obtained spectacular footage. The Bolsey was carried on an F-powered rocket and its lens looked out through a hole in the side of the body tube.

In 1970, Estes introduced the Cineroc, an aerial motion picture camera that used Super 8 cartridges to produce color film. The lens was mounted on the side of a 9.9" plastic cylinder housing the camera. A mirror set in a projecting hood provided a view looking toward the ground over the fins of the Cineroc's two-stage carrier.

The Cineroc was designed by Mike Dorffler, who had originally conceived the camera in 1965. The 3-ounce gadget featured a 10 mm focal length at f 11, shutter speed of 1/500 second, and an exposure rate of 30 frames per second. The rapid exposure rate created a slight slow motion effect. The cartridge contained 10 feet of 8 mm film which provided about 40 seconds of projection time.

The camera's battery-driven motor was started just before launch, and the ensuing launch, staging, and parachute ejection, all set against the rapidly receding ground below, made for fascinating footage — a viewer got the impression that he was actually riding on the rocket. (A variation in the shroud line arrangement could produce additional footage of the ground as viewed from the slowly descending rocket after ejection.)

Unfortunately, the Cineroc went the way of the Camroc in the 1970s, al-

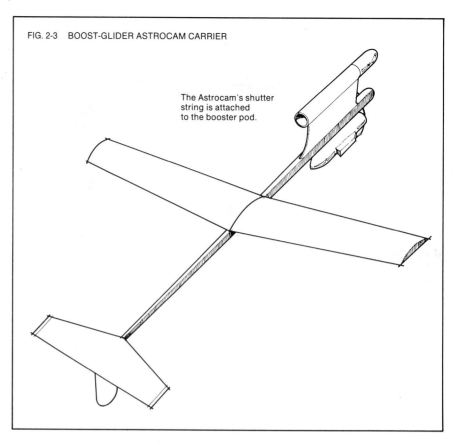

FIG. 2-3   BOOST-GLIDER ASTROCAM CARRIER

The Astrocam's shutter string is attached to the booster pod.

though Estes had a liberal supply of carrier rockets for the Camroc and Cineroc and advertised them in the catalog for several years after discontinuing the cameras.

Camrocs and Cinerocs can still occasionally be found in the hands of model rocketeers or among the forgotten stock in the back room of a hobby shop. Both cameras are collector's items.

Astrocam 110

Almost three years after production of the Camroc ceased, Estes introduced the Astrocam 110. Featured on the cover of the 1979 catalog, the Astrocam, designed by Mike Dorffler, uses 110 color print film in 12-exposure cartridges, but otherwise borrows liberally from both the Camroc and Cineroc. Shutter activation and aiming methods are the same as for the Camroc. The lens is side-mounted and a first surface mirror provides a view of the ground below; a light lock over the shutter prevents accidental exposure.

One important difference between the Camroc and the Astrocam is the fact that the shutter is activated by a spring rather than a rubber band. The shutter speed is now 1/500 second, although it was 1/1000 on some early Astrocams. As with the Camroc, a fast film is necessary; Estes recommends ASA 400. Because the film must be manually advanced the camera takes only one photo per flight.

The Astrocam's focal length is 30 mm at f 16. Fully loaded, it weighs 1.76 ounces. It is 6.5" long, 1.39" in diameter.

Estes sells the Astrocam as a package with the Delta II booster rocket, which is designed to fly with standard B or C engines and which has plastic fins and a prefinished white body tube for ease of assembly. Estes also offers the Challenger II, a D-powered booster with plastic fins and prefinished body, for high-altitude flights.

Flying the Astrocam

Flying the Astrocam is fairly simple. Once you've loaded the film in the camera and advanced it to the first frame, mount the camera on the booster with the light lock closed and the shutter string pulled per the instructions. Then mount the booster on the launcher, connect the igniter leads, and open the light lock. Now go through the preflight checklist in the Astrocam manual. Any rocketeer who has flown the Astrocam without using the checklist will tell you that he has lost a photo and wasted an engine because of something simple like forgetting to open the light lock.

Close the light lock and advance the film immediately after recovering the Astrocam. Make a practice of this and you won't have to worry about double exposures.

Estes recommends the C6-7 engine for vertical shots. The C6-7 works well, but you might find that the rocket is traveling a little too fast at ejection, which can cause parachute problems. In that case, try a C6-5. With a C6-5 the shorter delay will give you a vertical shot most of the time and the rocket

(Left) The Estes Camroc of 1965 used disks of Tri-X black-and-white film that required special processing and that yielded a circular print 3″ in diameter. (Right) The Estes Cineroc of 1970 is no longer in production but a few remain in the hands of collectors. The Astrocam 110 is physically similar to the Cineroc; both were designed by Mike Dorffler.

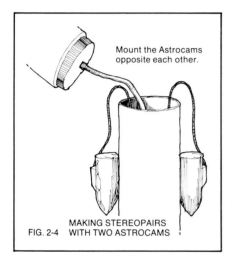

Mount the Astrocams opposite each other.

FIG. 2-4 MAKING STEREOPAIRS WITH TWO ASTROCAMS

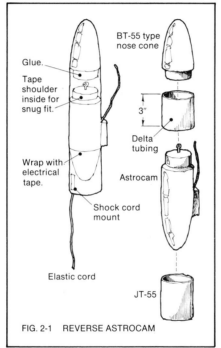

BT-55 type nose cone

Glue.

Tape shoulder inside for snug fit.

Wrap with electrical tape.

Delta tubing

3″

Astrocam

Shock cord mount

Elastic cord

JT-55

FIG. 2-1 REVERSE ASTROCAM

will be moving much slower at ejection.

The C6-5 engine is recommended for horizon shots, but a C6-3 will also provide a horizon shot about 75 percent of the time.

B engines work well for low-level Astrocam flights provided they are of the B8 or older B14 variety. Because the Astrocam vehicle weighs a bit more than the average rocket of its size, you must use an engine with a high average impulse to give the rocket an extra kick at liftoff. A B engine with a 3-second delay will normally give you a horizon shot, while a 5-second delay will provide a vertical shot.

For D-powered flights with the Challenger II, Estes recommends a D12-11 to obtain vertical shots. I've found that a D12-7 will also yield a vertical shot, though from a higher altitude. A D12-5 will usually provide a horizon shot, while a D12-3 will give you a sky shot.

Reverse Astrocam

Figure 2-1 shows a modification devised by Herb Desind that lets the Astrocam take photos while ascending. The "Delta Tubing" listed is Centuri Series 13 tubing. All other part numbers are Estes.

The shutter string must be lengthened and passed through the slit at the base of the Astrocam (now the top) and held in the cocked position by friction between the body tube of the booster and the piece of JT-55 on the Astrocam's nose.

The Astrocam will now take its photo while coasting upward after thrust burnout. B engines work best with this configuration and 3- or 5-second delays are ideal. If you wish to use a C engine, you should use a 5- or 7-second delay to permit the rocket to coast upward and slow down before ejection and shutter activation; this will help reduce the chances of the parachute fouling.

In addition to providing lower-altitude shots than are normally possible, reverse Astrocam shots can provide quite a bit of interesting information. In the accompanying photo, for example, you can see the ejection gases, along with the shock cord, coming out from beneath the camera. This gives you an idea of just how quickly the parachute is ejected, because the parachute is directly behind the shock cord and has already started to exit the body tube. (The escaping gases tell us that the parachute is already out.)

Other Astrocam techniques

For high-altitude Astrocam flights using engines more powerful than a D, you have two alternatives. You can strap the Astrocam to the outside of a high-power rocket and run the shutter string up to the host vehicle's nose cone or you can construct your own high-power launch vehicle. Figure 2-2 is a plan for an easy-to-build high-power Astrocam carrier.

If you are interested in building your own booster for the Astrocam, be aware that the special body tube used to accommodate the base of the camera was available from Centuri in 18″ lengths as the Series 13, part number 6010 ST1318. Centuri also stocked the plastic fin unit, part number 5467 F413L.

An Astrocam can also be carried aloft aboard a boost-glider carefully designed to accommodate this payload, Fig. 2-3. Shutter activation can be radio controlled, using either a mechanical linkage to release the shutter string or a circuit which fires a flashbulb, the flashbulb in turn burning through a section of the shutter string.

Flying tips

If you are flying in windy conditions, be sure to angle your launchpad into the wind to ensure that the Astrocam and its booster drift back into the recovery area. You should also cut a small hole in the center of the booster's parachute to increase the descent rate.

If you fear the shock cords might break, devise a way to recover the camera and the booster on separate parachutes. You might also consider using streamer recovery for ultrahigh flights; the Astrocam can withstand quite a bit

In this reverse Astrocam shot, the parachute has just begun to leave the body tube.

Using a C6-3 or C6-5 engine will yield a horizon shot on most Astrocam flights.

of landing impact. Finally, whatever recovery system you use, build it extra strong. The camera and booster will be traveling upward or downward at a high speed when the parachute or other recovery system is deployed and this puts quite a strain on the system. It is a good idea to use a nylon parachute and to epoxy the shroud lines onto the chute. You should also use a double shock cord.

If you want to be even more certain that your Astrocam returns to the recovery area, experiment with a drogue chute system. This involves first ejecting a small parachute or streamer to slow the camera and booster slightly (not enough for a landing) and to ensure that the larger chute will deploy properly. A timer, such as one of those formerly sold by CNA, then releases the larger chute, which slows the camera and booster enough for a safe landing.

When you fly your Astrocam is almost as important as how, due to varying light conditions. In general, late spring, summer, and early fall are the best times to fly photo missions because the earth's albedo (reflectivity) is then at its highest, thanks to the less oblique angle of the sun during these seasons. The best times of day are from 8 a.m. to 11 a.m. and from 1 p.m. to 4 p.m. when light levels are ideal and there are fewer long shadows.

Stereopairs

With two Astrocams, you can shoot stereopairs. A stereopair consists of two photos taken only a few degrees apart showing the same subject. When viewed side by side at close range, the stereopair provides a three-dimensional image, which can be helpful for comparing the relative sizes of objects in the photos.

The best method for making stereopairs is simply to strap both Astrocams onto the side of a large rocket, Fig. 2-4.

Aerial photo interpretation

The meaning of some information on aerial photos is obvious, but much more is not. For example, it is possible to determine the time of day by judging the lengths of the shadows on the ground. If there are no shadows, or very small shadows, the photo was made at noon or thereabouts. It is also possible to determine if a now fallow field has been used to grow crops; traces of furrows that are not apparent at ground level will tell you this. In the same manner, you can find evidence of old roads or buildings which no longer exist.

Partially overlapping Astrocam photos can form the basis of a striking photomosaic. To ensure a consistent scale, each photo should be shot from very nearly the same altitude.

Remember that your photos will be reversed, because pictures made by the Astrocam are taken through a mirror. Take this fact into consideration during any photo interpretation (unless you have been able to have the prints reversed during processing).

Some less than obvious information can be gleaned from Astrocam photos by studying color patterns in the same way that geographers, agronomists, and other scientists examine satellite photos. If you compare a photo of a field of corn, for example, with another photo showing the normal color of a field of corn, a difference in color between the two might indicate that the corn is diseased.

Varying coloration of bodies of water can likewise yield information. Shallow water is usually a lighter color than deep water (and sometimes blue as compared to green) and dark green indicates plant or algae growth. Dark brown shows that an area is disturbed by the turbulence of a heavy current or by runoff from another body of water.

H = OF ÷ I

Using the simple equation $H = OF \div I$, you can determine the altitude at which an Astrocam photo was taken. The variables are H = height of camera above the surface, O = size of an object on the ground shown in the photo, I = size of the same object's image as measured on the negative, and F = focal length of the camera.

The size of the object (O) should be measured in meters and the size of the object's image on the negative (I) should be in millimeters.

Let's make a trial run, substituting numbers in the equation. The focal length of the Astrocam (F) is a constant: 30. If we have a wall that is 5 meters long as measured on the ground and 4 millimeters long as measured on the negative, then the altitude of the Astrocam when the photo was made was $H = (5 \times 30) \div 4$, or 37.5 meters (the result is always in meters).

If you know the altitude from which an Astrocam photo was taken you can determine the size of an object on the ground by transposing the formula to $O = HI \div F$.

If we know that the Astrocam was at an altitude of 37.5 meters when the photo was taken and that the size of the object in question as measured on the negative is 4 millimeters, then $O = 37.5 \times (4 \div 30)$ or 5 meters.

Chapter 12 contains a computer program for this equation written in Level II Basic for the TRS-80. The equation is also readily adaptable to programmable calculators.

Astrocam photos frequently reveal features such as mowing, plowing, and drainage patterns that are not evident from ground level.

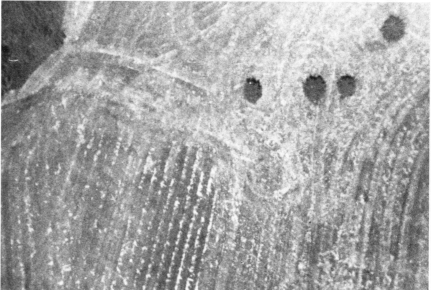

3. Contest-winning scale model rockets

By Thomas Hoelle

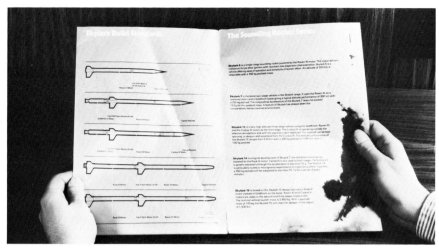

The British Aerospace Skylark is a one-, two-, or three-stage solid-propellant sounding rocket that uses a Raven XI, Goldfinch II, or Rook III as the main motor. Skylarks can carry loads ranging from 100 to 400 kilograms to altitudes of 150 to 1,500 kilometers. These photos show the front cover and an inside spread of a British Aerospace brochure on the Skylark.

In this chapter I'm going to assume that you've mastered basic and intermediate model building techniques, have acquired a reasonably complete set of modeling tools and materials, and are looking for new challenges. I'll present strategies you can use to build and fly contest-winning scale model rockets.

The nine steps

There are nine steps in building a contest-winning scale model.

1. Look at each of the seven scale events, study their requirements, and decide which event you'll enter.

2. Study several full-size rockets (which I'll call prototypes) that might be appropriate subjects for the event you've chosen. Choose one of these prototypes to model, taking into consideration the availability of data about the prototype, the degree of difficulty in constructing and flying the model, the time that will be required to build it, and (most importantly) your own likes and dislikes. Decide what scale the model will be.

3. Begin gathering data on the prototype — this includes photos, scale drawings, lists of specifications and dimensions, and any other material that might be helpful.

4. Organize the data in a notebook, prepare plans and working drawings in the scale you've chosen, decide what materials you'll use to build the model, and devise a step-by-step construction sequence.

5. Build and fly one or more rough-and-ready "boiler plate" models to verify the accuracy of your plans and the feasibility of your construction methods. If the boiler plate model is unstable, plan to add nose weight, clear plastic fins, or other devices to ensure flight stability in the scale model.

6. Build the actual model by sub-

assemblies. Build several of each sub-assembly, incorporating only the best version of each in the final model.

7. Prepare the model for the color coats of paint, then apply the color coats and any appropriate lettering, numerals, insignia, or other markings.

8. Prepare a scale documentation package of drawings, photos, and text to help the judges when they examine your model.

9. Enter your model in a contest, fly it, and take home a prize!

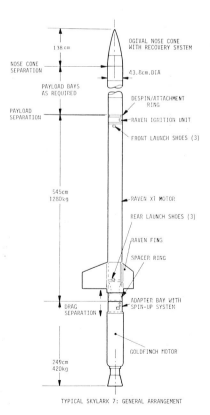

General-arrangement drawings are a big help during the early steps of designing a scale model rocket.

The seven events

Model rocket competitions in the United States are governed by the National Association of Rocketry (NAR). International contests are conducted according to rules established by the Federation Aeronautique International (FAI), the world governing body for all full-size and model sporting aviation activities, including rocketry. There are five NAR scale events; the FAI has two scale events — International Scale and International Scale Altitude.

The U. S. rules appear in the NAR's *United States Model Rocket Sporting Code*. This rule book, often called the "pink book" because of its pink cover, describes all competitive model rocket events recognized by the NAR. The FAI, which certifies all international records in flying-related sports, also publishes a sporting code which contains rules for the two international events I just mentioned — these rules are very similar to those for the NAR events.

Within each rule book, there are rules common to all scale events, Table 3-1, as well as rules that set apart each event according to its purpose and judging criteria, Table 3-2.

In Scale, Scale Altitude, Super Scale, International Scale, and International Scale Altitude the model is closely judged for craftsmanship as well as dimensional accuracy. Models entered in Scale Altitude, Space Systems, and International Scale Altitude are tracked for height and awarded points based either on maximum altitude or the smallest percentage of error from the altitude predicted by the contestant. A scale model of the prototype's launcher is required in Super Scale and is optional in Space Systems. The problems of an actual full-size prototype launch are simulated in Space Systems.

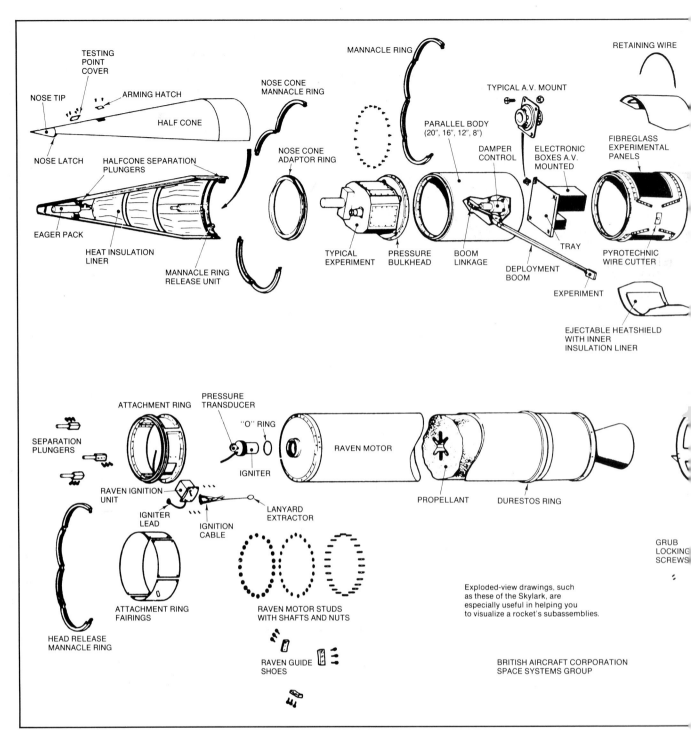

TESTING POINT COVER
NOSE TIP
ARMING HATCH
HALF CONE
NOSE LATCH
HALFCONE SEPARATION PLUNGERS
EAGER PACK
HEAT INSULATION LINER
MANNACLE RING RELEASE UNIT
NOSE CONE MANNACLE RING
NOSE CONE ADAPTOR RING
MANNACLE RING
TYPICAL A.V. MOUNT
RETAINING WIRE
PARALLEL BODY (20", 16", 12", 8")
DAMPER CONTROL
ELECTRONIC BOXES A.V. MOUNTED
FIBREGLASS EXPERIMENTAL PANELS
TYPICAL EXPERIMENT
PRESSURE BULKHEAD
BOOM LINKAGE
DEPLOYMENT BOOM
TRAY
PYROTECHNIC WIRE CUTTER
EXPERIMENT
EJECTABLE HEATSHIELD WITH INNER INSULATION LINER

ATTACHMENT RING
PRESSURE TRANSDUCER
"O" RING
IGNITER
RAVEN MOTOR
PROPELLANT
DURESTOS RING
SEPARATION PLUNGERS
RAVEN IGNITION UNIT
IGNITER LEAD
IGNITION CABLE
LANYARD EXTRACTOR
GRUB LOCKING SCREWS
ATTACHMENT RING FAIRINGS
RAVEN MOTOR STUDS WITH SHAFTS AND NUTS
HEAD RELEASE MANNACLE RING
RAVEN GUIDE SHOES

Exploded-view drawings, such as these of the Skylark, are especially useful in helping you to visualize a rocket's subassemblies.

BRITISH AIRCRAFT CORPORATION SPACE SYSTEMS GROUP

Table 3-3 breaks down point categories by event. Keep these categories in mind as you research and build your model, always concentrating on those areas that count the most in terms of points.

Choosing a prototype to model

After reviewing the scale events and choosing one in which to compete, the next step is to find one or more prototypes that match both the requirements of that event and your building abilities. There are many prototypes from which to choose, each calling for a different level of detail. A model of a Saturn V, for example, will require far more detail than a model of a Hawk or Honest John. Always consider if a scale model of a particular prototype will be a stable flier. Remember that a model rocket is not stabilized with gyroscopes or movable control surfaces, as are many full-size rockets and missiles, but with large rear fins.

The center of gravity of a model rocket should be at least one body diameter ahead of the model's center of pressure, which is usually located near the leading edge of the fins.

Generally, a model with good-sized fins that do not extend more than a third of the way from the base to the nose will be stable. You can, of course, add weight to the nose of a marginally stable model to move the center of gravity forward, but make sure your model will not require drastic alterations to achieve stability — such as 6 ounces of nose weight or large clear plastic fins. You can lose points for such changes.

You should also consider building a fairly simple model for your first scale model. After all, your first rocket was not a Saturn V or Space Shuttle, but a much simpler kit. You'll learn quite a bit from each scale model you build and can eventually work your way up to complicated models. Table 3-4 lists a few prototype possibilities by relative degree of difficulty. The first several prototypes listed are good choices for beginners.

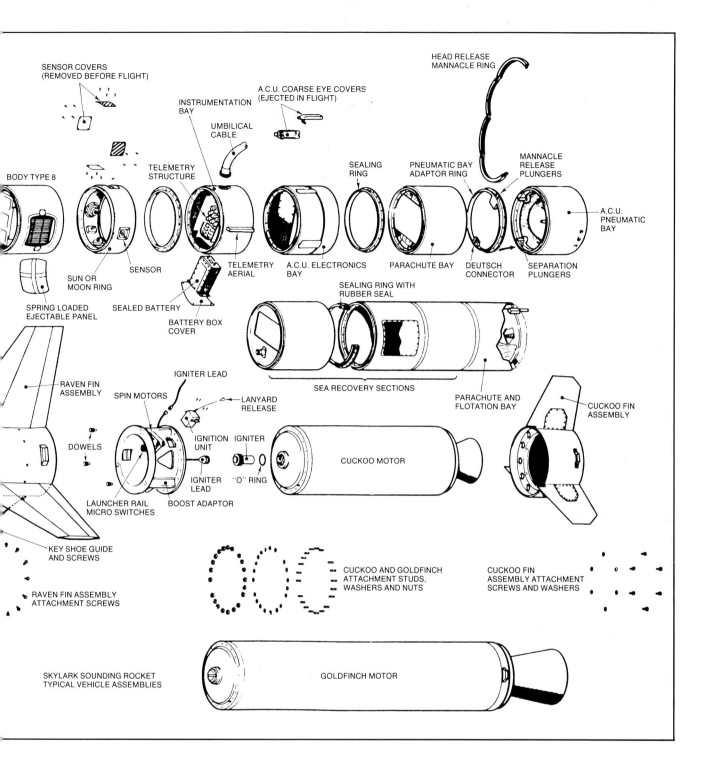

SENSOR COVERS
(REMOVED BEFORE FLIGHT)

HEAD RELEASE
MANNACLE RING

A.C.U. COARSE EYE COVERS
(EJECTED IN FLIGHT)

INSTRUMENTATION
BAY

UMBILICAL
CABLE

MANNACLE
RELEASE
PLUNGERS

BODY TYPE 8

TELEMETRY
STRUCTURE

SEALING
RING

PNEUMATIC BAY
ADAPTOR RING

A.C.U.
PNEUMATIC
BAY

SUN OR
MOON RING

SENSOR

TELEMETRY
AERIAL

A.C.U. ELECTRONICS
BAY

PARACHUTE BAY

DEUTSCH
CONNECTOR

SEPARATION
PLUNGERS

SPRING LOADED
EJECTABLE PANEL

SEALED BATTERY

BATTERY BOX
COVER

SEALING RING WITH
RUBBER SEAL

RAVEN FIN
ASSEMBLY

IGNITER LEAD

SPIN MOTORS

LANYARD
RELEASE

SEA RECOVERY SECTIONS

PARACHUTE AND
FLOTATION BAY

CUCKOO FIN
ASSEMBLY

DOWELS

IGNITION
UNIT

IGNITER

IGNITER
LEAD

"O" RING

CUCKOO MOTOR

LAUNCHER RAIL
MICRO SWITCHES

BOOST ADAPTOR

KEY SHOE GUIDE
AND SCREWS

CUCKOO AND GOLDFINCH
ATTACHMENT STUDS,
WASHERS AND NUTS

CUCKOO FIN
ASSEMBLY ATTACHMENT
SCREWS AND WASHERS

RAVEN FIN ASSEMBLY
ATTACHMENT SCREWS

SKYLARK SOUNDING ROCKET
TYPICAL VEHICLE ASSEMBLIES

GOLDFINCH MOTOR

If you can only obtain a little dimensional data for your prototype, consider building the model for Sport Scale or Space Systems.

Also take into account the amount of time you can spend on the project. The prototype you select if you have six months of building time is going to be quite different from the one you will choose if you have two months or less. In general, the simpler the prototype, the less time you will need. I'm frequently asked how long it took me to build a particular model; the answer is usually hundreds of hours.

From a competition standpoint, it is advisable to build models of rare or at least lesser-known prototypes. It may

be easier to obtain data about a common prototype, but if several other contestants choose the same rocket, the judges can group all of the models, compare them to one another, and easily pick the best.

As an additional consideration, select a prototype that has a sufficient amount of what I call "buildable detail." As a rule of thumb, the larger the model, the more details are needed. Big models without a sufficient amount of detail look bare and small blemishes stand out more, while the details on a small model are not that noticeable.

For example, suppose that you are constructing two Space Shuttle models in 1/72 scale and 1/200 scale. On the

1/72 version you would want to model each of the tiles on the underside of the Shuttle, while on the 1/200 model the tiles could merely be suggested by narrow paint lines. There is quite a bit of difference in the practicality of modeling a detail such as a tile when it is half an inch tall and when it is less than two-tenths of an inch tall! Obviously, you'll have more to show with a large model.

Having determined the approximate size and scale for a model, I then usually build around a commercially available body tube. This means that one body tube diameter determines the precise scale factor for the entire model. For example, my Ariane model is based

(Top) The U. S. Army's Honest John is a popular scale modeling subject because its long, narrow body and large fins make it inherently stable in flight. In addition, scale documentation data is easily available. (Right) The Little Joe II has also been popular. (Below) All models entered in a scale contest must make a successful flight; few are as impressive on liftoff as a Saturn V.

TABLE 3-1 RULES COMMON TO ALL SCALE EVENTS

The following rules apply to the five scale events in the NAR "Pink Book" and to both scale events in Section 4B of the FAI "Sporting Code."

A. You may enter any model that is a true scale model of an existing or historical guided missile, rocket vehicle, or space vehicle.

B. Models of amateur rockets or missiles are excluded unless they are of obvious historical significance.

C. Models may not be of non-flying or inert prototypes; however, data from this type of prototype can be used if you can prove that the information also applies to the version you have modeled.

D. Plastic model kits converted to flying models may not be entered in scale events; however, parts from plastic model kits can be used if these parts are pointed out to the scale judges. (Converted plastic models can be entered in an event called Plastic Model.)

E. A model can be entered in either Scale or Sport Scale, but not both.

F. Models built from commercial scale model rocket kits are acceptable only if you provide scale data in addition to that which was included in the kit.

G. You must model one particular serial-numbered rocket. Scale data can be used from multiple rounds if you can prove the data applies to the round you are building.

H. If the prototype is multistaged, you must build all stages, but not all stages need be operational.

I. You can use transparent plastic fins to increase stability; these will be judged for craftsmanship along with the model.

J. Your NAR membership number must appear on the outside of the model in numerals that are at least one centimeter high.

K. Your model will be judged in flight-ready condition less engines and recovery devices. Nothing can be added to or removed from the model from the time it is judged until it is flown except the engines and recovery devices. Any clear plastic fins, pop lugs, or other removable external devices required to make the model flyable must be in place for judging.

L. The model must make a single safe, stable flight. Failure to do so will result in disqualification.

M. You may not catch any part of your model unless this is specifically permitted by the contest director. Otherwise, catching your model will result in disqualification.

Build a boiler plate model before committing yourself to any final design for a scale model rocket. The inset photo shows my boiler plate version of the Ariane. The large photo and the full-page photo on page 24 show the full-size Ariane on its launchpad at Kourou, French Guiana.

on the diameter of an Estes BT-101 body tube, which has an outside diameter of 3.938″. To this, I added the thickness of the paint I was using, multiplied by two, to determine the overall diameter of the model's main body. Dividing this diameter into the diameter of the main body gave me a scale factor of 1/37.6367 (that is, 1 inch on the model equals 37.6367 inches on the prototype).

If the prototype has three body diameters, you may need to scratchbuild the other two sizes of tube. In that case, you may want to build your model around the most difficult tube to make, using a commercial tube for this portion. Generally, this is the largest diameter on the prototype.

The size of certain vitally important details (say spin motors) may force you to choose the scale factor that will allow those details to be large enough to be represented realistically.

If your model has scale markings or decals that are only available in a certain size, you might want to select your scale based on their dimensions. For example, if there are flags on your prototype and you can only find one size and don't want to make them from scratch, you will want to build the entire model based on the proportion of the size of those flags to the size of the flags on the prototype.

You may simply decide to build all of your models in a single scale, say 1/100, and gradually form a handsome collection that could become the basis of a striking massed display.

Whichever scale you select, the level of detail should be consistent throughout the model. If you have extremely detailed information on the nose section but only general information on the remainder of the prototype, don't detail the nose cone more than the rest of the model.

Obtaining scale data

There are many types of scale data, and you should try to get anything you can lay your hands on. Anything! But to be more specific, you should look for:
- Blueprints or scale drawings.
- Black-and-white and color photos.
- User's manuals.
- Public relations and sales material from the manufacturer.
- Books and magazine articles.

Based on the scale event you have selected, make sure that you can obtain at least the minimum amount of scale data, both dimensional and pictorial, to adequately document your model.

Where do you find such data? There is a surprisingly wide range of sources. Begin by looking at your public library for books and magazines that deal with space flight and military matters. For example, each year's edition of *Jane's All the World's Aircraft* is an excellent source of information. *Aviation Week*

TABLE 3-2 RULES UNIQUE TO EACH SCALE EVENT

A. SCALE
1. The purpose of this event is to produce an accurate, flying replica of a real rocket vehicle that exhibits maximum craftsmanship in construction, finish, and flight performance.
2. Data must be supplied to substantiate the model's adherence to scale in dimensions, shape, color, and paint pattern.
3. The event is scored by adding the static judging points (900 maximum) to the flight points (100 maximum).

B. SCALE ALTITUDE
1. The purpose of this event is to produce an accurate, flying replica of a real rocket vehicle that exhibits maximum craftsmanship in construction, finish, and flight performance, and to achieve the greatest possible altitude with the model.
2. Data must be supplied to substantiate the model's adherence to scale in dimensions, shape, color, and paint pattern.
3. Entries are judged and awarded scale points using the same criteria as Scale.
4. An entry may be disqualified if the scale qualities have been grossly subordinated in favor of increased altitude characteristics.
5. The event is scored by adding the scale points awarded (1,000 maximum) to the entry plus the altitude achieved in meters.
6. Scale Altitude competition is divided into the following classes based on engine size:

Engine Class	Total Impulse Range (Newton-seconds)	Maximum Weight (Grams)
¼A	0.00-.625	60
½A	.626-1.25	60
A	1.26-2.50	60
B	2.51-5.00	60
C	5.01-10.00	120
D	10.01-20.00	180
E	20.01-40.00	240
F	40.01-80.00	453

C. SUPER SCALE
1. The purpose of this event is to produce an accurate flying replica of a real rocket vehicle and an accurate working replica of its accurate launching complex, both of which exhibit maximum craftsmanship in construction, finish, and performance.
2. Data must be supplied to substantiate the model's adherence to scale in dimensions, shape, color, and paint pattern.
3. The event is scored by adding the scale points awarded to the model (1,000 maximum) to the scale points awarded to the launching complex (1,000 maximum).

D. SPORT SCALE
1. The purpose of this event is to produce a flying replica of a real rocket vehicle that exhibits maximum craftsmanship in construction, finish, and flight performance. Sport Scale differs from Scale in that the dimensions of the model are not measured and that the model is not inspected at close range by the judges.
2. Data must be supplied to substantiate the model's adherence to scale in shape, color, and paint pattern but not dimension.
3. The model will be inspected by the judges from a distance of at least one meter.
4. Sport Scale is scored by adding the static judging points (800 maximum) to the flight points (200 maximum).

E. SPACE SYSTEMS
1. The purpose of this event is to duplicate in miniature the situation and problems encountered in the full operation of a sounding rocket or space vehicle with instrumented range support. The contestant acts as a project officer and is faced with the same decisions and must make many of the same compromises as his professional counterpart. As in full-scale operation, he does not have control over all of the factors that will influence the flight and the success of his vehicle.
2. Entries are judged and awarded scale points for the prototype minus launcher using the same criteria as for Sport Scale.
3. Each contestant may optionally enter a model of the prototype's launcher. If entered, the launcher is judged using the same standards as those applied to the model.
4. Each entry must have its own electrical ignition system and carry a standard NAR one-ounce payload.
5. The Range Safety Officer will assign a launch time, designate a safe recovery area, and assign launching areas for your model.
6. You must predict the altitude that your model will reach and the model is tracked for altitude.
7. Rescheduling the launch window costs 25 points and you are penalized 100 points if the model is not launched within the window set by the Range Safety Officer.
8. 250 points are lost if the model does not land within the safe recovery area.
9. The winner is determined by multiplying the sum of the static score of the model and the static score of the optional launcher by .75, adding the model's altitude, and then subtracting the percentage of deviation from predicted altitude as well as subtracting any penalties incurred.

F. INTERNATIONAL SCALE (CLASS S7)
1. The purpose of this event is to produce an accurate, flying replica of a real rocket vehicle that exhibits maximum craftsmanship in construction, finish, and flight performance.
2. Data must be supplied to substantiate the model's adherence to scale in dimensions, shape, color, and paint pattern.
3. The event is scored by adding the static judging points (900 maximum) to the flight points (100 maximum).
4. Each entry must make a stable flight. Two opportunities will be available for this purpose, time and weather permitting.

G. INTERNATIONAL SCALE ALTITUDE (CLASS S5)
1. The purpose of this event is to produce an accurate, flying replica of a real rocket vehicle that exhibits maximum craftsmanship in construction, finish, and flight performance, and to achieve the greatest possible altitude with the model.
2. Data must be supplied to substantiate the model's adherence to scale in dimensions, shape, color, and paint pattern.
3. Entries are judged and awarded scale points using the same criteria as Scale.
4. An entry may be disqualified if the scale qualities have been grossly subordinated in favor of increased altitude characteristics.
5. The winner of this event is determined by adding the scale points awarded (1,000 maximum) to the entry plus the altitude achieved in meters.
6. Each entry must make a stable flight. Two opportunities will be available for this purpose, time and weather permitting.
7. Scale Altitude is divided into the following classes based on engine size:

Class	Total impulse range (Newton-sec)	Engine class	Maximum weight (Grams)
S5A	0.00- 2.50	A	60
S5B	2.51- 5.00	B	90
S5C	5.01-10.00	C	120
S5D	10.01-40.00	D-E	240
S5F	40.01-80.00	F	500

TABLE 3-3   POINT CATEGORIES BY EVENT

Event	Scale	Scale Altitude	Super Scale	Sport Scale	Space Systems	Int'l Scale	Int'l Scale Altitude
ROCKET							
A. Scale data	50	50	50	400	400	50	50
B. Accuracy of major dimensions	200	200	200			350	350
C. Accuracy of color and markings	100	100	100				
D. Accuracy of details	50	50	50				
E. Craftsmanship	300	300	300	300	300	300	300
F. Degree of difficulty	200	200	200	100	100	200	200
G. Flight	100	100	100	180	180	100	100
H. Mission				20	20		
LAUNCHER							
A. Scale data			50				
B. Accuracy of major dimensions			200				
C. Accuracy of color and markings			50				
D. Accuracy of details			50				
E. General appearance			100				
F. Craftsmanship			250				
G. Degree of difficulty			200		50		
H. Operations			100				
For international events, accuracy is defined as follows:							
A. Scale appearance						50	50
B. Body and nose cone						100	100
C. Color and markings						100	100
D. Fins						100	100

and Space Technology is the finest magazine in its field. *Aviation Week and Space Technology* also publishes an important annual issue that includes descriptions of missiles currently in use and that contains a long list of rocket manufacturers and other contractors, with addresses.

The government document section at your library will have a NASA index titled *Scientific and Technical Aerospace Reports (STAR)*, which includes descriptions of NASA articles and reports about rocketry. Once you have obtained the number of the NASA publication dealing with your prototype from *STAR*, you can locate the article or report in your library, or order it from either the Clearinghouse for Federal Scientific and Technical Information or the Superintendent of Documents in Washington, D. C.

Of course, there are always back issues of model rocket magazines to search for scale data. *Model Rocketry* published a good deal of information on scale models and you may be able to obtain back issues from fellow hobbyists. NAR's *American Spacemodeling* (formerly *The Model Rocketeer*) frequently runs articles and photos useful to scale modelers.

I have found that information from one source often leads me to new sources. For example, when I wrote to

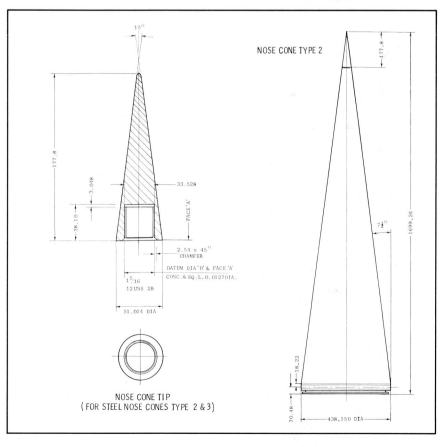

If you can obtain even more detailed drawings of subassemblies, so much the better. Here are the Type 2 nose cone and the nose cone tip for the Skylark.

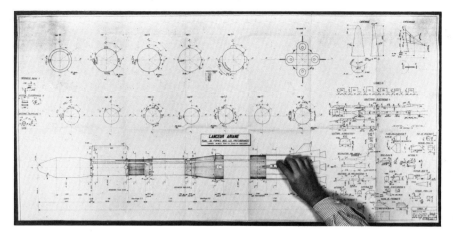

the European Space Agency for information on the Ariane satellite launching rocket, I received not only the material I requested, but the name and address of another source of information on the Ariane. A letter to British Aerospace about the Skylark sounding rockets yielded a number of helpful drawings as well as a brochure with several color photos.

Some blueprints have given me leads in the form of the names of contractors written in the information block (usually found in the lower right-hand corner of the print). When doing research for my Trailblazer II model, I found the name of the project manager on the blueprint and he proved to be a good source. Blueprints often contain identification numbers for other drawings you may want to order.

Model rocketry club newsletters often feature articles on a particular prototype. Also, if you write to newsletter editors, they will usually provide the names and addresses of their contacts for information. Friends who have collected data on a prototype you wish to model are, of course, an invaluable resource; they've already done half of the research!

You can also try tracking down a hobbyist who has already built a model of your prototype; he may help you avoid mistakes.

Writing to a contact

Once you have established your primary information sources among manufacturers and the like, sit down and establish just what you want to request and how you are going to ask for it. When writing a contact for the first time, be brief and to the point — let him know who you are, what your background in model rocketry is, what project you have in mind, and explain exactly what information you desire (photos, dimensional and color data, blueprints, or whatever). Also, to lend creditability, include a copy of the rules for your scale event with the letter.

Do not overload your contact with voluminous requests for data or you may be refused. Keep in mind that the per-

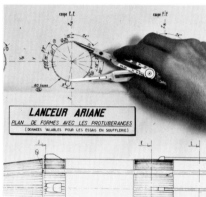

(Top) Large drawings such as these for the Ariane may contain almost all the information you require. (Above) Proportional dividers can make quick work of transferring dimensions from large drawings to your assembly drawings.

son you are writing to has many job duties, among which answering your correspondence is a minor concern.

Once you have received a reply, you'll be able to judge the value of the contact by the amount of information sent, as well as by the tone of the letter. If the contact seems favorably disposed, you will be in a position to write again for more information.

I have found that it's best to ask for a little information several times rather than ask for a lot in one letter. Again, you must remember that your requests are not the only concern of the person you are writing; something that requires a small amount of effort, such as finding and sending one blueprint, will most often be given attention before a more involved request.

Boiler plate models

It is important to build a rough version of the vehicle, a boiler plate model, in order to assess the flight characteristics of your scale model.

With a boiler plate model you can get a good idea as to the altitude your model will achieve and can experiment with various engines in order to build for the best performance. It helps to know how straight the model will fly, and whether it will require nose weight

or clear plastic fins. In addition, the boiler plate version will give you a good idea about the approximate weight of the scale model.

This is very important if you are pressing against the upper weight limit. For example, if your boiler plate model weighs around 15 ounces, be extremely weight conscious when building the scale model.

Planning the construction sequence

Always take the time to plan the entire construction sequence for your model, from initial sizing of major components through the final details of finishing. Many hobbyists omit this step, and I neglected it myself when I first began to build scale models. I would let my models just "fall together" but ran into time-wasting problems when they did not fall together the way I had hoped. It is a good feeling to have a schedule and to stick to it, knowing what the next step in the building process is going to be.

Part of the planning process is keeping a notebook. In the notebook should be descriptions of potential problems and their solutions (as you discover them), ideas for making details, and drawings of individual parts with their

TABLE 3-5 MINIMUM SCALE DOCUMENTATION BY EVENT

SCALE

1. Overall length
2. Any tubing diameters
3. Nose cone length
4. Fin length, width, and thickness
5. Length of any transition pieces
6. Color pattern documentation, consisting either of photographs or text
7. One clear photograph, halftone, or photocopy
8. A table of all required dimensions for both the prototype and the model

SCALE ALTITUDE

Same as Scale.

SUPER SCALE

1. Minimum scale documentation for the model is the same as for Scale.
2. The requirements for the launching complex are:
   a. Scale factor
   b. Color pattern documentation consisting of photographs or text
   c. One clear photograph, halftone, or photocopy
   d. Substantiation of the fact that the particular launcher modeled was indeed used to launch the prototype
3. A table of all required dimensions for both the prototype and the scale model

SPORT SCALE

1. To substantiate similarity of outline, either:
   a. A line, tone, or color drawing
   b. One or more clear photographs, halftones, or photocopies sufficient to show the outline and general configuration of the prototype
2. To substantiate finish, color, and markings, either:
   a. One or more clear photographs, halftones, or photocopies, including at least one in color
   b. Other published pictorial representations such as a color painting or a drawing from a magazine
   c. A detailed written description from a reliable source of the color scheme and markings, accompanied by a drawing of the prototype showing the color scheme

SPACE SYSTEMS

Minimum scale documentation for both the model and the launcher is the same as for Sport Scale.

INTERNATIONAL SCALE

The modeler must provide dimensional data including the length and diameter of the prototype from any reliable source, as well as one photograph of the prototype.

INTERNATIONAL SCALE ALTITUDE

Minimum scale documentation is the same as International Scale.

---

dimensions. You may even want to record the number of hours you've spent on the project.

It is important to think of ways to make your job easier. Making jigs and fixtures is one way to ensure accuracy when fabricating parts and attaching these to your model.

I experiment with various techniques when constructing details. Many times, I have to discard the product of one technique, but sooner or later I'll come upon just the right method, sometimes accidentally. You should be continually thinking of new ways to solve detailing problems.

Form the habit of reading a variety of modeling books and magazines, not just those concerned with rocketry — publications about plastic modeling, flying model airplanes, and model railroading often contain information on materials and techniques you can adopt.

The importance of building by subassemblies

The best approach is to view the model as a series of subassemblies. Using the Ariane as an example, the major subassemblies on my model included the first stage, the second and third stage, and the payload section. Within each subassembly, there are a number of details. For instance, the first-stage subassembly contains the fins, body tube, fin shrouds, acceleration motor covers, engine fuel lines, four Viking IV engines, and markings and graphics.

I first isolated and defined the individual elements, or details, of each subassembly and built these one at a time. However, when the same details were required for more than one subassembly, I made all of these at one time, thereby ensuring consistency among these parts and saving time as well.

I build at least three copies of each subassembly. The first is a rough version to see if the overall concept is correct. With the second copy, I begin to concentrate on the details, making these as accurate as possible. The third copy usually goes into the model — it's based on everything I've learned working on the two earlier versions. Generally, the second-best subassemblies become parts of a back-up model.

Remember: Plan out your work, keep a notebook, and read modeling books and magazines in search of techniques that will help you build your model faster and more efficiently. Build by subassemblies and ruthlessly discard or find other uses for inferior parts. Incorporate only your best work into the final model.

Allowing for paint thickness

If you were to measure the diameter of a tubular part on the prototype you're modeling, what would you actually be measuring? The answer is the tube and two thicknesses of paint. The measurements taken by the judges will include the diameter of the building materials and the thicknesses of the primer, color coats, and any protective clear varnish. (They will measure a body tube section of the same type used to build your model, then measure your finished model and subtract the smaller measurement from the larger to obtain the paint thickness.) As a result, you must take the thickness of the paint on the prototype into account if you want full points for dimensional accuracy. Keep in mind that you will be dealing with two thicknesses of paint, one on each side of the body tube.

In order to accurately compensate for the thickness of your paint, paint several test pieces of plastic with a representative sample of the finish you will be using. Measure the plastic before and after painting with a micrometer (the difference between the two measurements will, of course, be the thickness of the paint). Try to duplicate the thickness of the paint on your finished model by using the same spraying or brushing techniques you used on the test pieces.

To illustrate the importance of this, if I hadn't taken paint thickness into account when calculating the dimensions for the fins of the Ariane, I could have lost up to 20 points in the judging of that one area.

Recovery techniques

Two types of material can be used for parachutes on scale model rockets — silk and aluminized Mylar. If you are using silk chutes, make them large enough (or use two) to ensure a gentle landing for your model. The lines should be nylon cord in order to minimize the possibility of ripping and you may want to run the shroud lines up over the canopy for greater strength. If you use Mylar, which is lighter than silk and sometimes more visible, I recommend .5-mil thickness.

If the model is to separate into two pieces, it is important that your chutes be packed one inside the other. Pack the chute that stays with the lower half of the model inside the chute for the upper half. This way, the chute for the upper half will extract the chute for the lower half as the model separates. If you use more than one chute but the model stays in one piece, pack the chutes separately. Be sure to use enough recovery wadding to prevent damage to the chutes. You may also want to loosely wrap the chutes in paper or wadding.

Should your model separate completely into two sections or be strung together with elastic shock cord? If you are flying a Saturn V with only the Apollo capsule for a nose cone, your choice will not make a lot of difference. If, however, your Saturn V separates at the first and second stage adapter, the decision will make a difference. Generally, you risk swing damage when a model comes down as one unit, especially if the two sections are large, as with the Saturn V.

Therefore, two large sections should be brought down separately unless you can connect them with a long enough piece of monofilament fishing line and elastic so that one hangs well below the other during descent. You should remember, too, that if one chute out of two fails on a single recovery unit, you will have a fair chance of a safe recovery, while if one chute of a two-unit setup fails, you are going to lose half of your model. If in doubt, try both ways with a boiler plate model.

Scale documentation package

Prepare a logically organized package of materials that will help the judges determine the accuracy of your model. This package, most often in the form of a loose-leaf notebook or ring binder, will contain some but not all of the material you gathered while researching the prototype.

It should contain:
● A brief history of the prototype, including its applications.
● A table stating the major dimensions on the prototype and the model.
● A list of the reference materials you used and their sources.
● Important drawings and blueprints.
● Photographs of the prototype. Black-and-white or color prints are best but halftones clipped from magazines are usually acceptable.
● Data to prove the accuracy of the color scheme and markings on your model.

In addition, you should provide information about how you built the model, including:
● A list of which parts were scratch-built and which were obtained from commercial sources.
● Notes on how you solved difficult steps in assembling and finishing the model.
● Statements telling the judges any facts about the model that aren't obvious from viewing it. For example, if your nose cone was turned from balsa and everyone thinks it looks like plastic because the grain is completely concealed, mention it. The judges may overlook an extremely difficult detail unless it is pointed out.

The information should be presented in the order in which it will be judged (if the dimensions are judged first, dimensional data should be placed at the front of your scale pack). Include an index or table of contents, as well. The rules state that a judge may take off points if your documentation package complicates judging. Table 3-5 summarizes the amount and type of documentation required for each event.

The competition flight

After static judging comes the most exciting part of scale modeling — the actual flight of the model in competition. The flight of your model can be critical; I have seen many a scale event decided by a model's flight performance. Placing first in static competition isn't enough — a bad flight can cause disqualification.

Pick the type of engine that gave the best flights with your boiler plate model and be sure to static test at least one engine from each package. To ensure the most consistent performance, it's best to use engines from the same batch or production run.

Wear white gloves when you handle the model to prevent fingerprints and body oils from smudging it.

Finally, be sure that the model is recovered in a safe manner. Make certain that no one catches it unless the "no catching" rule is suspended by the Contest Director.

Homemade launch control systems are among the items most frequently sold or bartered at swap meets. Some of the systems feature integrated circuit timers and light-emitting diodes that provide a visual countdown to zero.

4. Launch control systems

By David Babulski

Advances in model rocket ground support electronics have closely followed developments in solid-state electronics. In the early days, model rocketeers were content with simple manual DC switching devices to provide electrical power for engine ignition, Fig. 4-1. Because early igniters were made from Nichrome wire, a large amount of current was needed to heat the wire sufficiently to ignite the propellant in the rocket engine. A lead-acid car battery was the most common power source.

This technology worked well until rocketeers began clustering engines. Several igniters were connected in parallel so that all the engines in the cluster would ignite at the same time. At first, the answer to the need for a larger supply of electrical current was to use a bigger battery. Unfortunately, this proved to be very cumbersome, particularly for portable launchers. Because substantial current was lost in the wires from the launch controller to the launcher, a way of shortening the distance between the ignition battery and the igniter was needed.

This gave rise to the next advance in ground support technology — the relay-controlled launch system. This system allowed the ignition battery to be placed close to the igniter while permitting the rocketeer to remain a safe distance from the launcher, Fig. 4-2.

A relay is nothing more than a switch operated by an electromagnet. A small battery in the hand-held portion of the launch controller supplies the operating voltage for the relay, which is located at the launcher. When the firing button is depressed, the relay actuates. A set of the relay's contacts close, completing the circuit between the ignition battery and the igniter.

This was an improvement over the earlier system, but was not without its problems. Relays are electromechanical devices, so they are subject to both electrical and mechanical failure, usually at the most inopportune moment. Relay systems are relatively expensive

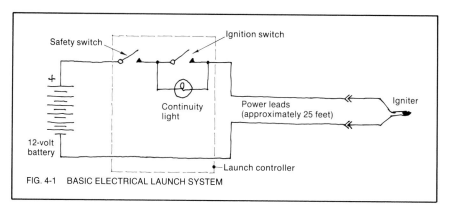

FIG. 4-1   BASIC ELECTRICAL LAUNCH SYSTEM

Safety switch

Ignition switch

+

Continuity light

Power leads (approximately 25 feet)

Igniter

12-volt battery

Launch controller

to operate, as well, because two batteries are required — one for the relay and one for the igniter. The slow rate at which ignition occurs is also a drawback.

The problems with relay-controlled launch systems led to the next development — the capacitive-discharge system, Fig. 4-3.

In general terms, a capacitor is an electronic device designed to store an electric charge and release it when needed. When used in a launch controller, the capacitor stores a large amount of power from the ignition battery and releases it through the igniter in a sudden pulse, ensuring fast, reliable ignition of single and clustered engines. With capacitive-discharge systems, the ignition battery can be reduced in size, lowering cost and increasing portability compared with earlier methods.

Capacitive-discharge systems were the accepted standard until the Estes Solar igniter and flashbulb ignition came on the scene.

The Solar igniter can be successfully ignited with the current from four small 1.5-V AA alkaline dry cells connected in series. As a result, the simple DC controller made a big comeback.

The simple DC controller could not be used for flashbulb ignition, however, because hobbyists soon discovered that the small amount of current in the circuit during a continuity check would set off the flashbulb. The search was on for a flashbulb-safe launch controller. The answer was found in the solid-state launch controller, Fig. 4-4.

With the solid-state launch controller, a transistor is used as a switch to complete the circuit between the ignition battery and the flashbulb. A tiny idle current through the transistor is used with a meter or piezoelectric alarm as a flashbulb-safe continuity check.

All kinds of variations have followed. The solid-state controller is frequently used in conjunction with the capacitive-discharge system and multi-pad systems have been developed with a separate transistor switch for each launchpad.

When low-cost integrated circuits became readily available, the automatic digital launch controller was the next innovation, Fig. 4-5. When the start switch is closed momentarily on the digital launch controller, the 555 IC timer generates one voltage pulse per second. These voltage pulses cause the digital circuitry to count down from 9 to 0. The MAN-7 is a light-emitting diode matrix which displays the count. When a count of 0 is reached, the 74190 provides a voltage pulse to the 2N2222 transistor. The transistor then acts as a closed switch to send current through the igniter. Thus, the digital launch controller is really a combination of digital electronics with a solid-state launch controller.

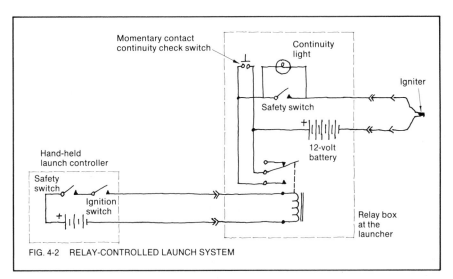

FIG. 4-2   RELAY-CONTROLLED LAUNCH SYSTEM

The widespread use of Estes Solar igniters (far right), which require far less current than earlier types, has made possible simple 6-V launch controllers that operate from low-current batteries. For example, the Estes Solar Launch Controller (above) operates on four 1.5-V alkaline AA cells; the Power-Pulse launch controller (right) uses a small 6-V Polaroid Polapulse P-100 battery.

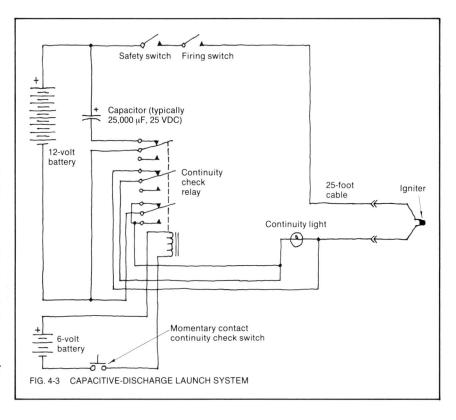

FIG. 4-3   CAPACITIVE-DISCHARGE LAUNCH SYSTEM

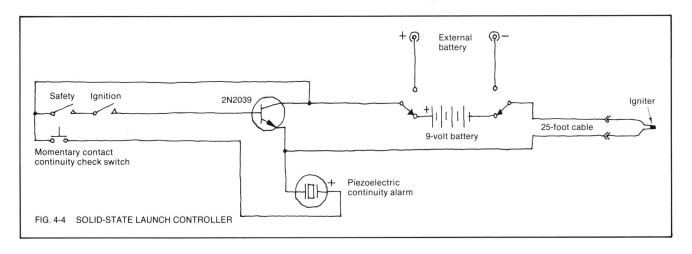

FIG. 4-4 SOLID-STATE LAUNCH CONTROLLER

Safety   Ignition
2N2039
External battery
+   −
Igniter
9-volt battery
25-foot cable
Momentary contact
continuity check switch
Piezoelectric
continuity alarm

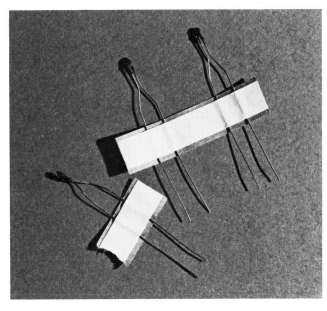

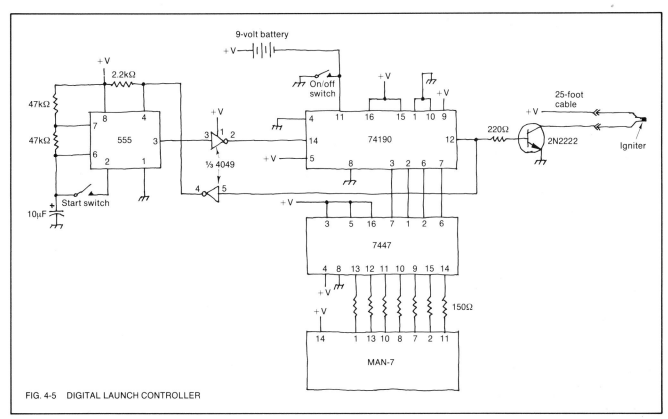

FIG. 4-5 DIGITAL LAUNCH CONTROLLER

9-volt battery

47kΩ
47kΩ
2.2kΩ
+V
On/off switch
+V
+V
25-foot cable
+V
555
8 4
7
6 2 1
3
⅓ 4049
3 1 2
4 5
74190
4 11 16 15 1 10 9
14
5
8 3 2 6 7
220Ω
2N2222
Igniter
Start switch
10µF
7447
3 5 16 7 1 2 6
4 8 13 12 11 10 9 15 14
150Ω
MAN-7
14 1 13 10 8 7 2 11

5. A countdown timer for model rockets

By Jack Cunkelman

The circuit in the schematic drawing, Fig. 5-1, is designed to provide an automatic countdown and launch for model rockets. It features an on/off switch, an arm/safe switch, and a launch (run) button. Light-emitting diodes indicate continuity and provide a visual countdown. When the launch sequence is started, nine LEDs flash one after another, counting down from 9 to 0.

The timer uses four easy-to-find integrated circuits — a 555 timer, 7400 quad NAND gate, 74193 4-bit up/down counter, and a 74154 4 to 16 decoder.

The countdown timer uses the TTL logic family. The LS series logic family is functionally identical, but uses 80 percent less power. The voltage to these chips must not exceed 5.25 V. The function of D1 is to drop the 6 V from the power supply battery to approximately 5.13 V. This is the drop across a silicon diode junction.

The 555 is connected as an astable timer. The output pulse on pin 3 is connected to one input of the 7400 AND gate. Pin 2 is high and the pulses from the 555 are gated through. When the "0" LED is lit, pin 2 is pulled low and the pulses are blocked. This stops the countdown and acts as a control gate.

Pulses from the control gate are fed into pin 5, which is the up counter. The down pin (4) input is tied to +5. It is used as an up counter. It is a 4-bit counter, so it can count to 16. We are only using a part of the count; we are really counting up, but it doesn't matter — it's all in how the LEDs are arranged. Pin 14 is the clear input; when this pin is high, the counter is set to 0. When it is grounded, the counter is started. The output of the chip is a 4-bit binary word which is fed to the decoder chip.

The decoder chip takes the 4-bit binary word at its input and makes an output low. All of the others remain high. This provides a path for the LEDs to ground and they light. The 330-ohm resistor is used for current limiting.

The SCR is Radio Shack 276-1067, rated for 6 A, 200 V. The SCR requires a positive pulse to latch it on. The 7400 feeding the gate is used as an inverter. It takes the grounding transition and makes it a positive pulse to turn on the SCR. The SCR will remain on until the supply voltage is cut off. This happens when the igniter burns and opens the circuit.

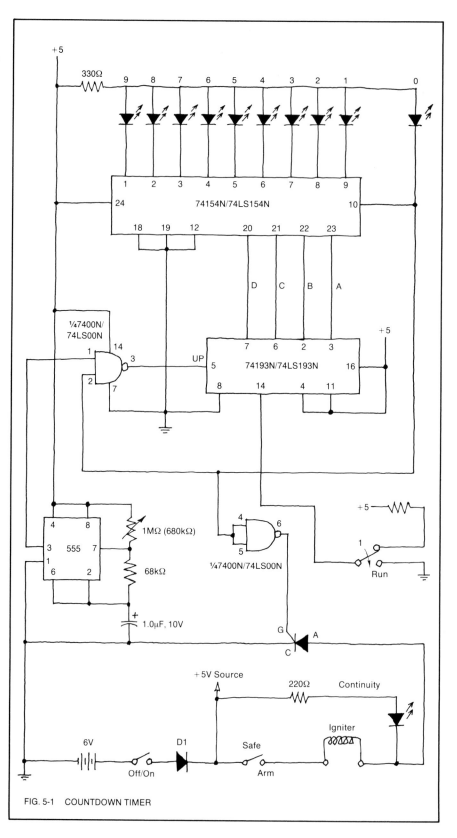

FIG. 5-1   COUNTDOWN TIMER

6. Radio telemetry transmitters

By David Babulski

Have you ever asked yourself, while watching your latest bird climb majestically into the atmosphere, "I wonder what it's like up there?" For over 20 years, some model rocketeers have been using telemetry to answer that question.

Telemetry (literally "measurement from far away") is a branch of engineering concerned with electronic systems for gathering information (such as temperature, pressure, humidity, or acceleration in one or more axes) from inaccessible locations (such as your rocket at 2,000 feet), and transmitting it to a base station where it is received, recorded, and analyzed. Figure 6-1 is a simplified diagram of a typical system.

The first model rocketry telemetry system was flown by John Roe and Bill Robson at NARAM-2 in August 1960. Since then, several manufacturers have supplied telemetry transmitters — Estes offered the Transroc and Competition Model Rockets has sold the Foxmitter.

The sending portion of a basic telemetry system consists of a radio transmitter connected to a transducer (also known as a sensor). The transducer generates or modifies an electrical signal in response to some external physical stimulus. The electrical signal, which is an analog of the physical change sensed by the transducer, is supplied to the radio transmitter.

The transmitter's job is to impose (modulate) this audio frequency signal onto a radio frequency carrier which is radiated by an antenna. A receiver at the base station tuned to the same radio frequency as the transmitter demodulates the carrier and decodes the sensor's information, converting it back into an audio frequency current which is fed to an instrument such as a tape recorder or chart recorder. Once the information is recorded, it can be analyzed at any time, obtaining answers to your question "what's it like up there?"

Although NASA spends millions of dollars on its telemetry equipment, you can gather useful information at little cost. Ground equipment can be as simple as an ordinary CB receiver and a portable tape recorder.

You can build your own transmitter or assemble a Foxmitter. G. Harry Stine's classic *Handbook of Model Rocketry* (Follett Publishing Company, Chicago, 1976) contains schematics and construction details for a small 27-MHz transmitter (which I'll call Stine's transmitter) and the next chapter tells how to convert a Radio Shack radio control module to a telemetry transmitter. No matter which design you choose, keep several factors in mind.

● You are limited to a total liftoff weight of 453 grams, so the transmitter and sensors must be lightweight.

● The Federal Communications Commission allows a maximum power output of 100 mW for unlicensed transmitters (Part 15, Subsections 15.116 to 15.141, FCC regulations).

● The FCC limits antenna length for unlicensed transmitters to 36".

● Two bands are commonly used for low-power, unlicensed telemetry operations: 26.99 to 27.26 MHz and 49.82 to 49.90 MHz. The transmitter must maintain a frequency tolerance of .01 per-cent, so it must be controlled by a crystal.

Basic definition of a transmitter

Reduced to its simplest form, a radio transmitter, telemetry or otherwise, is nothing more than an alternating current generator. When graphed with reference to time, an alternating current can be represented by a sine wave, Fig. 6-2. It is the job of the transmitter to change direct current, usually supplied by a battery, to alternating current at a specific frequency. Frequency refers to how fast the AC voltage alternates between positive and negative potentials in one second. The unit of frequency is the Hertz (Hz). Thus, a sine wave that alternates between maximum positive and negative values at a rate of 1,000 times per second would have a frequency of 1,000 Hz.

If the frequency of the alternating current generated by the transmitter is in the radio frequency (RF) portion of the electromagnetic spectrum, an antenna connected to the output of the transmitter will radiate the sine wave as an electromagnetic wave. This radiated RF energy is called a carrier, an appropriate term because you can modify, or modulate, the carrier to convey information from the transmitter to the receiver.

A typical model rocket transmitter can be divided into four basic sections: the RF section, modulator, sensor, and power supply, Fig. 6-3. The power supply is usually simply a small 9- to 22.5-V battery. The other three sections are more complex, so let's examine them separately.

RF section

As in most small radio telemetry

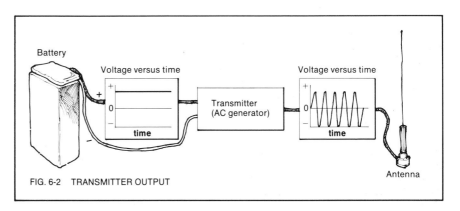

FIG. 6-1   SIMPLIFIED DIAGRAM OF A TELEMETRY SYSTEM

FIG. 6-2   TRANSMITTER OUTPUT

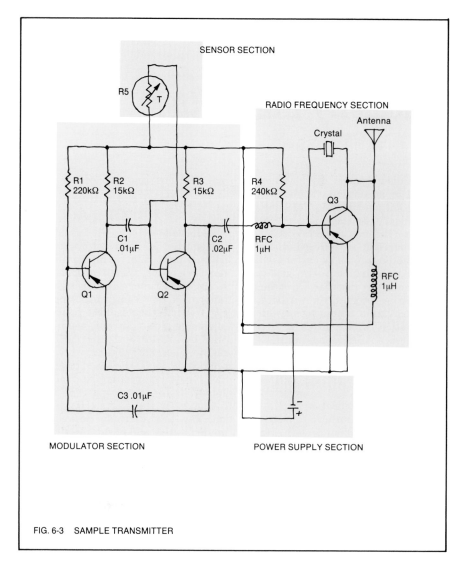

SENSOR SECTION

RADIO FREQUENCY SECTION

Antenna

Crystal

R5 T

R1 220kΩ R2 15kΩ R3 15kΩ R4 240kΩ

Q3

C1 .01µF C2 .02µF RFC 1µH

RFC 1µH

Q1 Q2

C3 .01µF

MODULATOR SECTION

POWER SUPPLY SECTION

FIG. 6-3 SAMPLE TRANSMITTER

transmitters, the RF portion of our sample transmitter in Fig. 6-3 is a crystal-controlled oscillator.

An oscillator converts direct current from the battery to alternating current at specific frequencies. Two types of crystal-controlled oscillator designs are commonly used in radio telemetry applications: Colpitts and Pierce. Our sample transmitter, the Estes Transroc, the Foxmitter, and Stine's transmitter,

all use the Pierce oscillator. The transistor — Q3 in our sample transmitter — serves as the active element in this conversion process. The crystal acts as a regulator to ensure that the frequency of the AC produced by the oscillator is maintained within strict tolerances.

In some cases it is necessary to amplify the RF energy produced by the oscillator in order to produce sufficient

output from the transmitter. Another reason for amplifying the output of the oscillator is to isolate the oscillator from changes in antenna loading, as might happen if the antenna whipped in flight. Load changes could cause the oscillator output to vary in frequency.

Two types of RF amplifiers are often used in radio telemetry applications: common base and common emitter, Fig. 6-4. Figure 6-5 is the schematic of the Estes Transroc, which used a final amplifier in the RF section. Transistor Q6 is the RF amplifier in a common base configuration.

Modulator

By itself, all the RF section does is generate a radio frequency carrier. To send useful information, the carrier must be modified in some way. This process is controlled by the modulator.

A modulator can be as simple as a switch to turn the carrier on and off (this is called carrier wave or CW modulation) or, as in Fig. 6-3, a second oscillator operating in the audio frequency (AF) range can be used. The modulator in our sample transmitter is called a multivibrator. Transistors Q1 and Q2 work in conjunction with the resistors and capacitors in the circuit to generate a square wave, Fig. 6-6.

The square wave output from the modulator is applied to the base of transistor Q3 and causes Q3 to vary the amount of RF energy that is generated. Notice that the amplitude, or vertical size, of the sine wave generated by the oscillator varies in step with the output of the modulator; this is amplitude modulation (AM) and is the form of modulation most commonly used with model rocket radio telemetry transmitters.

Be careful when using modulated oscillators because if the amplitude of the signal is too great the oscillator may be pulled off its correct frequency and you may lose data (you may also cause interference with another radio service). If the transmitter includes an RF amplifier, it is much safer and more efficient to modulate the amplifier rather than the oscillator.

The Estes Transroc used a slightly different type of modulator. In the telemetry mode, the Transroc was modulated with a modified CW technique. In Fig. 6-5, Q1 is a unijunction transistor which works with the resistors and capacitors in the circuit to generate a series of pulses. Transistor Q3 acts as a switch. The pulses generated by Q1 switch Q3 on and off; Q3 in turn controls the on and off time of the transmitter. Information is conveyed by varying the on time versus the off time of the transmitter. Transistor Q4 acts as a current regulator to control the amount of RF energy amplified by Q6. Transistor Q5 acts as the oscillator.

The Transroc used an antenna match-

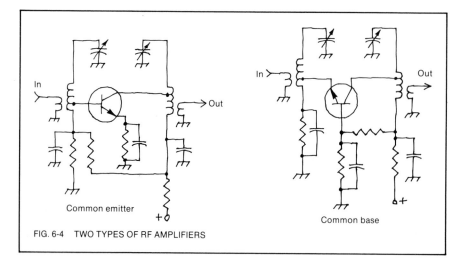

In

Out

Common emitter

+

In

Out

Common base

+

FIG. 6-4 TWO TYPES OF RF AMPLIFIERS

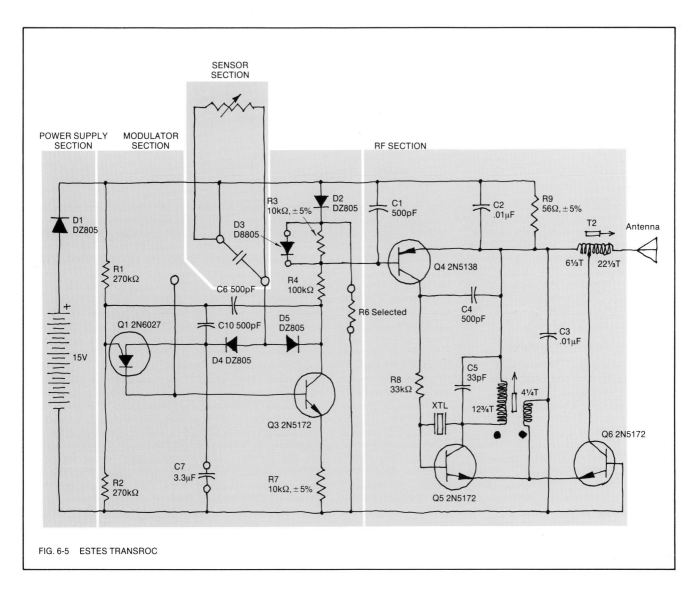

FIG. 6-5 ESTES TRANSROC

ing coil, T2. Our sample transmitter, Fig. 6-3, does not use this component. An antenna matching coil is valuable because it allows for maximum energy transfer to the antenna; it also helps to further isolate the oscillator from variations in antenna loading.

Sensor

In our sample transmitter, the sensor is a thermistor. A thermistor's electrical resistance decreases as temperature rises; these variations are used to change the audio frequency signal produced by the modulator.

Resistive sensors are available for measuring variables other than temperature. Also, by modifying the modulator, you can use inductive or capacitive sensors. Thus, the Foxmitter can be adapted to use an inductive sensor to measure changes in acceleration.

You can find more information on radio technology in *The Radio Amateur's Handbook*, published by the American Radio Relay League. (Do not write Estes Industries regarding the Transroc — it has been out of production for some time and Estes can no longer provide information about it.)

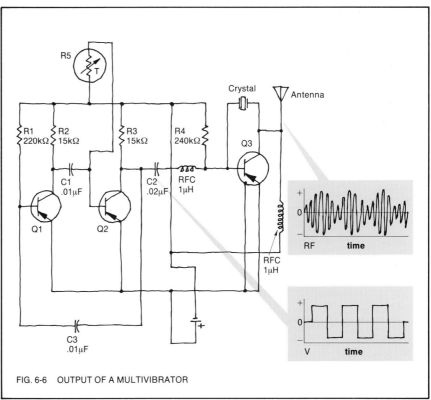

FIG. 6-6 OUTPUT OF A MULTIVIBRATOR

7. Building and flying a two-channel telemetry package

By David Babulski

Radio Shack sells a pair of inexpensive, prewired radio control modules (277-1012) designed for toys, models, and other devices. I'll first describe how to convert the transmitter module to model rocket telemetry applications (the other half of the pair is a receiver module that is not used in this project) and then explain how to install the transmitter in a model rocket.

Modulator changes

The Radio Shack transmitter consists of three basic sections: a modified Colpitts crystal-controlled oscillator, a common-emitter RF amplifier, and a two-transistor multivibrator modulator, Fig. 7-1. The transmitter operates on 27.145 MHz; its output is a little less than 100 milliwatts. All components are mounted on a $\frac{7}{16}$" x $3\frac{11}{16}$" printed circuit board. After the sensors are added, the transmitter weighs 50 grams, or 1.75 ounces.

The stock modulator is designed for use with manually actuated switches. Well, you can't manually throw switches when the transmitter is 2,000 feet above ground! My modifications electronically simulate the manually oper-ated switches. I'll describe the changes required to telemeter air temperature and roll rate, but keep in mind that other variables can be measured.

The modulator, Fig. 7-2, is a two-transistor, astable (also called free-running) multivibrator circuit consisting of transistors Q3 and Q4, diodes D1 and D2, resistors R6 through R11, and capacitors C9 and C10. The rate at which capacitors C9 and C10 charge and discharge through resistor pairs R7-R10 and R8-R9 causes this circuit to produce a square wave with a frequency of either 500 or 3,000 Hz.

Which frequency is produced depends upon which resistor pairs are connected to positive supply voltage. If positive voltage is connected to R7-R10, the frequency will be 500 Hz. In this case, diodes D1 and D2 act as open switches to remove R8-R9 from the circuit. If positive voltage is applied to the anodes of D1 and D2 they act as closed switches to add R8-R9 to the circuit. The combination of resistor pairs R8-R9 and R7-R10 causes the circuit to produce a frequency of 3,000 Hz.

The modifications are adding a circuit to control positive voltage to the resistor pairs and replacing R9 and R7 with resistive sensors (the temperature sensor is a thermistor, the roll rate detector is a cadmium sulfide photocell). One telemetry channel will consist of variations from the basic 500-Hz frequency; the other will consist of variations from the basic 3,000-Hz frequency.

The basic frequencies of the two channels are not far enough apart to eliminate the possibility of confusion when analyzing data. For example, if the sensor in the 500-Hz channel increased the frequency to 1,000 Hz and the sensor in the 3,000-Hz channel lowered its frequency to 1,000 Hz, it would be impossible to tell which was which if both signals were of equal duration. To avoid this confusion, the modulator control circuit turns on one channel for about ten times longer than the other channel so that you can easily tell them apart.

Modulator control circuit

The modulator control circuit, Fig. 7-3, is not particularly complex. A 555 integrated circuit timer, IC-1, is the active element in the circuit. The fixed and variable resistors work in conjunction

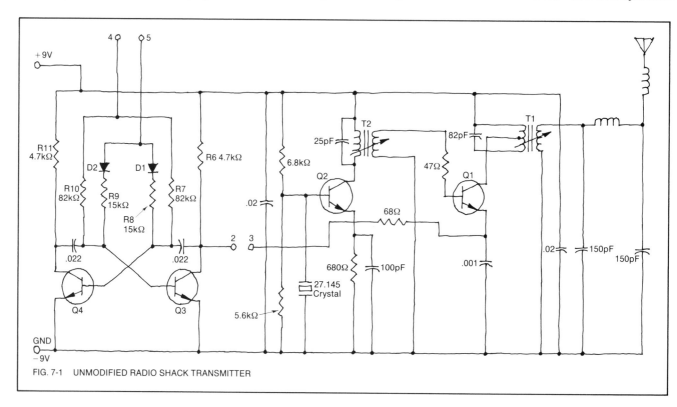

FIG. 7-1   UNMODIFIED RADIO SHACK TRANSMITTER

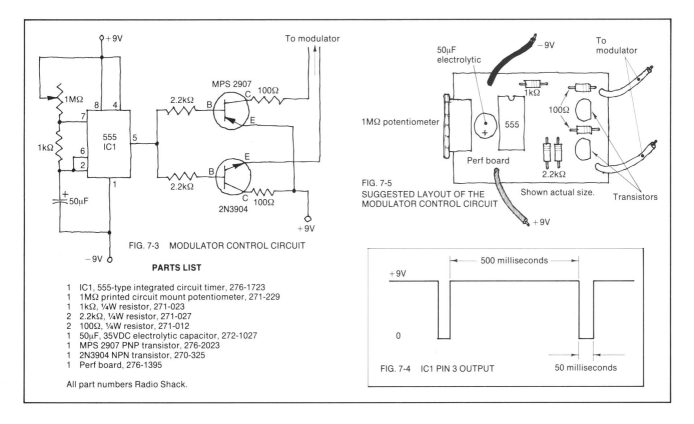

FIG. 7-3 MODULATOR CONTROL CIRCUIT

PARTS LIST

1 IC1, 555-type integrated circuit timer, 276-1723
1 1MΩ printed circuit mount potentiometer, 271-229
1 1kΩ, ¼W resistor, 271-023
2 2.2kΩ, ¼W resistor, 271-027
2 100Ω, ¼W resistor, 271-012
1 50µF, 35VDC electrolytic capacitor, 272-1027
1 MPS 2907 PNP transistor, 276-2023
1 2N3904 NPN transistor, 270-325
1 Perf board, 276-1395

All part numbers Radio Shack.

FIG. 7-5
SUGGESTED LAYOUT OF THE
MODULATOR CONTROL CIRCUIT

Shown actual size.

FIG. 7-4 IC1 PIN 3 OUTPUT

with the 50-microfarad electrolytic capacitor to control the frequency of a square wave produced at pin 3 of IC-1. The circuit is adjusted to produce an asymmetrical square wave at pin 3 of IC-1 consisting of a positive pulse of 500 milliseconds duration, Fig. 7-4.

This sequence of positive and negative pulses repeats as long as power is applied to the modulator control circuit. The pulse train produced by the modulator control circuit is applied to the bases of both an NPN and PNP transistor. These transistors act as switches.

Note that +9 volts is applied to the PNP emitter and the NPN collector. With no input to the bases, both transistors are "cut off" and act as open switches. When the output of the control circuit goes positive, the NPN transistor acts as a closed switch (the PNP transistor remains cut off) to supply +9 volts to the modulator. When the control circuit output goes negative, the PNP transistor now acts as a closed switch (the NPN transistor now remains cut off) to supply +9 volts to the modulator. The 2.2-kilohm and 100-ohm resistors act as current limiters to protect the transistors.

You must now decide which channel will be on longer than the other. A good rule of thumb is to have the channel you expect to change the most stay on the longest. In this application, temperature is expected to change more rapidly than roll rate. Thus, the output of the NPN transistor switch will be connected to the temperature resistor pair in the modulator while the output of the PNP transistor switch will be

connected to the roll rate resistor pair.

Modification instructions

The first step in modifying the transmitter is to build the modulator control circuit. The easiest way to build this circuit is to use perf board and point-to-point wiring. Figure 7-3 is the schematic and parts list for the modulator control circuit; Fig. 7-5 is a recommended parts layout. Although parts placement is not critical, the layout shown will make point-to-point wiring easier. Be sure to use a heat sink when soldering the leads of the integrated circuit transistor. Check your wiring job for errors, shorts, and unsoldered connections before applying power to

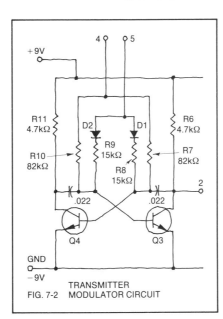

TRANSMITTER
FIG. 7-2 MODULATOR CIRCUIT

the circuit. When the modulator control circuit is complete, set it aside (you will test it a little later).

The remainder of the transmitter modifications can be completed in a series of 16 steps — the result is shown in Fig. 7-6 and Fig. 7-7 is the schematic of the assembled telemetry transmitter.

☐ 1. Solder a jumper between points 2 and 3 on the transmitter board.
☐ 2. Desolder and remove 82-kilohm resistor R7 from the transmitter board.
☐ 3. Cut two 4″ pieces of stranded hookup wire. Twist these wires together. Strip about ¼″ of insulation from the two wires on both ends of the twisted pair.
☐ 4. Solder the thermistor to the two wires on one end of the twisted pair.
☐ 5. Solder the other ends of the twisted pair to where R7 was removed. You have now replaced R7 with the temperature sensor.
☐ 6. Desolder and remove 15-kilohm resistor R9.
☐ 7. Repeat step 3.
☐ 8. Solder the photocell to the other two wires on one end of the twisted pair.
☐ 9. Solder the other ends of the twisted pair to where R9 was removed. You have now replaced R9 with the roll rate sensor.
☐ 10. Solder the wire from the output of the NPN transistor switch on the modulator control circuit to point 4 on the transmitter circuit board.
☐ 11. Solder the output wire from the PNP transistor switch on the modulator control circuit to point 5 on the transmitter circuit board.
☐ 12. Solder the RED wire from the 9-V

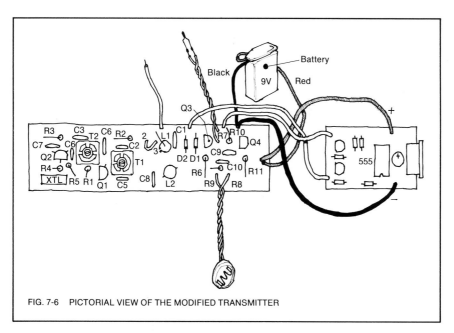

FIG. 7-6   PICTORIAL VIEW OF THE MODIFIED TRANSMITTER

battery connector to the 9V + point on the transmitter circuit board.

☐ 13. Solder the BLACK wire from the 9-V battery connector to the GND point on the transmitter circuit board.

☐ 14. Solder the + wire from the modulator control circuit to the 9V + point on the transmitter circuit board.

☐ 15. Solder the ground wire from the modulator control circuit to the GND point on the transmitter circuit board.

☐ 16. Solder a 24″ length of wire to the ANT point on the transmitter circuit board. This serves as a temporary antenna while you check out the transmitter.

Transmitter test

To perform an operational check on the transmitter you will need a CB receiver tuned to Channel 15. Turn on the CB receiver and connect a 9-V bat-

tery to the transmitter. Depending on where the potentiometer on the modulator control circuit is set, you will hear a single tone or two tones rapidly repeating. Adjust the potentiometer on the modulator control circuit so that one tone is on for about .5 second. The other tone will sound like a short tone burst.

If you have an oscilloscope, connect the oscilloscope probe to pin 3 of the 555 integrated circuit timer. Adjust the potentiometer on the modulator control circuit for a positive-going pulse of 500 milliseconds and a negative-going pulse of 50 milliseconds.

Next, check sensor operation. Touch the temperature sensor and verify that the received tone changes on the temperature channel. Cover the photocell and verify that the received tone changes on the roll rate channel.

If you do not receive any signal from the transmitter, recheck all connections and solder joints.

Installing the transmitter in a rocket

Keep a number of design factors in mind when you select a rocket to carry your telemetry package.

For example, you should decide — before you build your rocket — what altitudes you want to gather data from. In professional circles this is sometimes referred to as a "mission profile." With the mission profile in mind and knowing the weight of your telemetry package, you can make an initial estimate of how much total thrust you will need to meet your goals.

Determine if there is adequate recovery area at your flying site. If the recovery area is too small for your anticipated mission profile, you run the risk of losing the telemetry package.

Calculate how much you can afford to spend for a model and engines. This is particularly critical if you expect to conduct telemetry operations over an extended period of time. Costs have a nasty habit of rapidly getting out of hand.

How long do you want your telemetry package to remain airborne? The answer to this question (taking into account the size of the recovery area as well) will determine what type of recovery device to use.

When choosing a launch vehicle, be certain that the inside diameter of the body tube is large enough to house the telemetry package. Will major modifications be required to build a payload section?

Now make whatever trade-offs are required. For example, you may have to limit your mission profile if it looks like long-term expenses will be a prob-

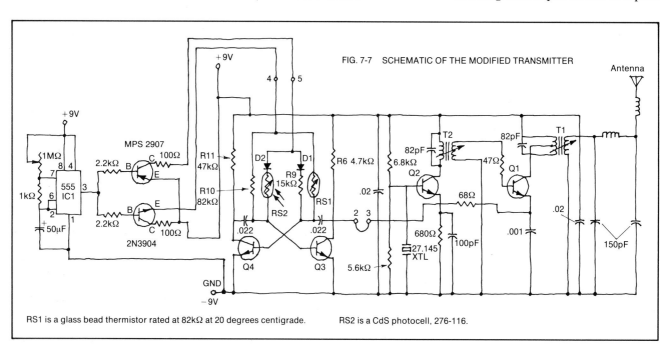

FIG. 7-7   SCHEMATIC OF THE MODIFIED TRANSMITTER

RS1 is a glass bead thermistor rated at 82kΩ at 20 degrees centigrade.   RS2 is a CdS photocell, 276-116.

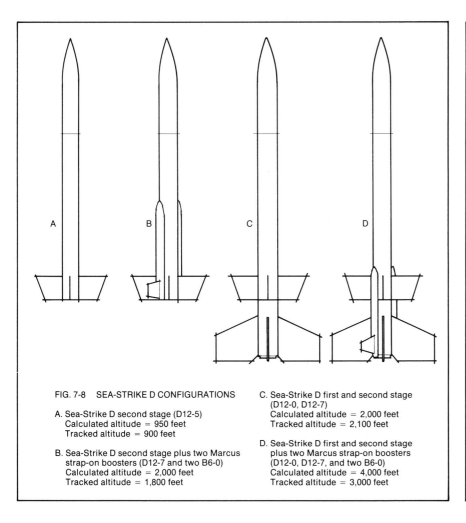

FIG. 7-8   SEA-STRIKE D CONFIGURATIONS

A. Sea-Strike D second stage (D12-5)
   Calculated altitude = 950 feet
   Tracked altitude = 900 feet

B. Sea-Strike D second stage plus two Marcus
   strap-on boosters (D12-7 and two B6-0)
   Calculated altitude = 2,000 feet
   Tracked altitude = 1,800 feet

C. Sea-Strike D first and second stage
   (D12-0, D12-7)
   Calculated altitude = 2,000 feet
   Tracked altitude = 2,100 feet

D. Sea-Strike D first and second stage
   plus two Marcus strap-on boosters
   (D12-0, D12-7, and two B6-0)
   Calculated altitude = 4,000 feet
   Tracked altitude = 3,000 feet

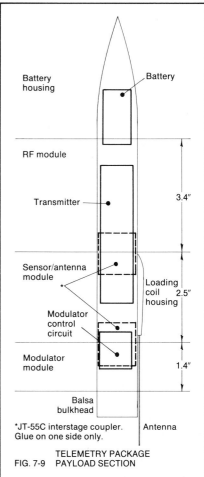

*JT-55C interstage coupler.
Glue on one side only.

TELEMETRY PACKAGE
FIG. 7-9   PAYLOAD SECTION

lem. If you have a new flying site with a larger recovery area, you may want to try for higher altitudes. This decision, in turn, will have to be weighed against increased model and engine costs.

Launch vehicle

I selected an Estes Sea-Strike D modified to use two Competition Model Rockets Marcus strap-on boosters as the launch vehicle for the telemetry package.

I wanted to achieve altitudes of 1,000 to 4,000 feet and my calculations indicated that various combinations of the Sea-Strike D and Marcus strap-on boosters, Fig. 7-8, would reach the upper and lower altitudes of my mission profile as well as those in between. Flight tests have shown fairly close agreement with the predicted values. The only exception is combination D; drag forces at the higher velocities were greater than expected and the rocket reached a lower altitude than I'd calculated.

Several modifications to the Sea-Strike D will increase its performance and reliability.

Begin by replacing the balsa second stage fins with 1/32" plywood. You can use one of the balsa fins as a template. The plywood fins should be sanded to a symmetrical shape (airfoil) and epoxied to the second stage body tube.

Next replace the 1/8" launch lugs supplied with the kit with 3/16" launch lugs. A "full house" Sea-Strike D with parallel staged boosters and the telemetry package weighs almost 220 grams and a 1/8" rod is just not rigid enough to ensure a stable launch.

Finally, cut mounting holes for the Marcus strap-on boosters, as described in the instructions included with the boosters.

Payload section

You'll need the following Estes parts to build a payload section for the Sea-Strike D: a 12-inch length of BT-55 body tube, an NB-55 balsa bulkhead, and a JT-55C interstage coupler.

The payload section is designed for maximum serviceability and reliability, Fig. 7-9. It is divided into four sections, or modules, which are fastened together with vinyl tape before launch.

You may ask "Why four sections? Why not use a single section of BT-55?" The answer is serviceability. By dividing the payload into four sections each area of the telemetry package can be serviced independently without having to remove the entire package.

The first section is the modulator module; it's a cover for the transmitter's modulator control circuit. A balsa bulkhead is mounted on the aft portion of this module and a JT-55C interstage

coupler is mounted to the forward part of the module. The balsa bulkhead is used to mate the payload section to the rocket. In addition, the payload recovery device is attached to the bulkhead. When this module is removed, the modulator control circuit is exposed for adjustments.

The second section is the sensor and antenna module. The telemetry sensors are mounted on the outside surface of this module. The antenna is mounted to the outside surface 90 degrees from the launch lugs and sensors, Fig. 7-10.

The transmitter is modified to make the antenna and sensors pluggable. Connectors small enough to fit in a model rocket telemetry package are hard to find. An inexpensive alternative is to make your own connectors from a 16-pin integrated circuit DIP socket (Radio Shack 276-1998) and a 16-pin DIP header (276-1980). Using a razor saw, you can cut off sections of both the header and the socket. The header portion becomes the male part of the connector and the socket portion becomes the female part. Use care when soldering to the leads of the socket; it is easily damaged by excessive heat.

The third module is a cover for the RF portion of the transmitter. The fourth module, the nose cone, serves as a housing for the transmitter battery.

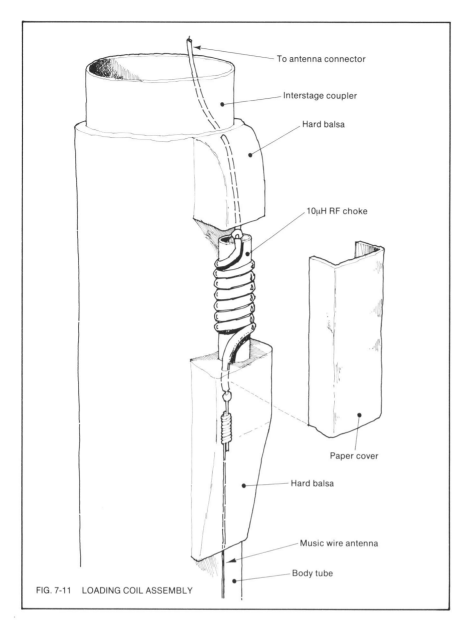

To antenna connector

Interstage coupler

Hard balsa

10μH RF choke

Paper cover

Hard balsa

Music wire antenna

Body tube

FIG. 7-11   LOADING COIL ASSEMBLY

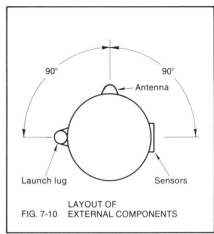

90°   90°

Antenna

Launch lug   Sensors

LAYOUT OF
FIG. 7-10   EXTERNAL COMPONENTS

Remove the bottom conical portion of the plastic Sea-Strike D nose cone and place the battery inside the cone.

Before you button up the payload section for flight, install small pieces of foam rubber between the circuit boards and between the battery and the transmitter circuit board. The foam rubber will provide protection against high g forces during launch and reduce the possibility of damage should the recovery system fail.

One of the most important parts of your transmitter installation is the antenna. Be sure to install the antenna correctly. The optimum length for the antenna for our transmitter on 27.145 MHz would be 5.52 meters, or 18.1 feet. You are limited by FCC regulations to an antenna not more than 91 centimeters, or 36 inches; a far cry from 18.1 feet! Because you are limited to a physical length of 36 inches, your only recourse is to change the antenna's electrical length. Do this by adding a loading coil at the base of the antenna.

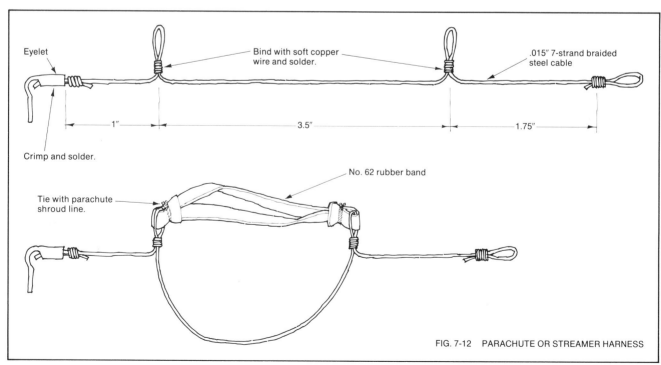

Eyelet

Bind with soft copper wire and solder.

.015″ 7-strand braided steel cable

Crimp and solder.

1″   3.5″   1.75″

Tie with parachute shroud line.

No. 62 rubber band

FIG. 7-12   PARACHUTE OR STREAMER HARNESS

(The transmitter already has an output loading coil, but installing a second coil makes the antenna more efficient.)

Because space inside the payload compartment is limited, the additional antenna loading coil mounts on the outside surface of the payload sensor and antenna module, Fig. 7-11. The coil is a 10-microhenry RF choke (such as Radio Shack 273-202, which may still be available) placed between two hard balsa mounting blocks and epoxied in place. A paper cover over the assembly will reduce drag. One lead of the RF choke is fed through a hole in the sensor and antenna module and is soldered to the antenna plug.

The RF choke's other lead is soldered to a 35″ length of .050″ music wire. Because this is a high-stress joint, bind the music wire to the RF choke lead with a short piece of soft copper wire before soldering. Music wire is used for the antenna to reduce whipping during ascent. If stranded wire were used as the antenna, it would flutter excessively, increasing drag and causing load variations at the transmitter.

Recovery considerations

I designed the payload parachute and streamer harness, Fig. 7-12, with maximum reliability in mind. First modify the balsa bulkhead by drilling a ¼″ hole through its center and gluing a ¼″ hardwood dowel into the hole. The parachute and streamer harness will then be fastened to the dowel with a ½″ wood screw through an eyelet on the harness.

Make the harness from a piece of .015″ 7-strand braided steel cable of the type used with control line model airplanes, forming loops as shown in Fig. 7-12. A No. 62 rubber band tied to both loops serves as a shock absorber. Attach the shroud lines from the parachute and streamer to the loop at the free end of the wire harness. (The wire harness is so short that it has no effect on the antenna's performance.)

You can arrange the recovery system so that the payload and rocket descend together by attaching the rubber shock cord supplied with the Sea-Strike D kit to the parachute or streamer harness. Mount the rubber shock cord to the rocket as shown in the kit instructions.

Flight considerations

Flying a telemetry package is a little different from flying a sport or competition model. It is very difficult to fly a telemetry model rocket by yourself and obtain good, usable data because you must know the altitude of the rocket at apogee. Your telemetry data is not much good if you don't know where it was measured. In most cases, therefore, one person mans the tracker and receiver; the other prepares the payload and launches the rocket.

Also, if you want to receive data from the transmitter during the entire flight, you must place the receiver at least a quarter of a mile away from the launcher because maximum energy is radiated at right angles to the transmitter antenna. If your receiver is located at the launcher you will receive the trans-mitter signal when the rocket is on the launchpad, but when the rocket rises a few hundred feet the energy radiated by the transmitter antenna will pass over your receiver antenna and you will lose the signal.

Some final comments

You may be tempted to use C6-0 engines in the Marcus strap-on boosters. This is not a good idea for two reasons.

First, the burn time of the C6-0 is very close to that of the D12-7. Flight tests with the second stage plus Marcus boosters fitted with C6-0 engines have revealed a tendency for the second stage to veer off from the desired path when the boosters separate from the second stage. I believe this is caused by slight differences in burn time between C6-0 engines, causing one booster to separate sooner than the other.

Second, when C6-0 engines are used in the Marcus boosters with the first and second stages of the Sea-Strike D (that is, a full house configuration), the center of gravity moves dangerously close to the center of pressure. This can result in marginally stable launches.

Flashbulb ignition is recommended when using the Sea-Strike D with Marcus boosters — all three engines must ignite at the same time.

The payload configuration I've just described also works well when the converted Radio Shack transmitter is used with the Flight Systems Maverick or Competition Model Rockets Chameleon, both F-100-powered electronic payloaders.

8. Using small rockets for atmospheric sampling

By Eric V. Nelson

Sampling particulates such as dust, pollen, and fungal or bacterial spores in the atmosphere often involves collecting these materials at ground stations over a period ranging from a day to several months. The resulting sample is then analyzed. Such techniques are simple and inexpensive but can only sample at ground level or at the maximum height of structures in the area.

Of course, particle traps can be attached to radio control model airplanes or to full-size aircraft but these methods are expensive and require skilled flight personnel; further, exhaust from the internal-combustion engines used may contaminate the sample.

Model rockets have been used in air pollution studies to determine the altitudes of temperature inversions. I have found that model rockets can also lift particle traps and thus overcome the disadvantages of both fixed trap sites and aircraft trap carriers. I developed a technique in which small rockets carried aloft agar-coated slides to trap microbiological air contaminants. The rockets used commercial parts, did not require skilled helpers, and allowed sampling at altitudes of up to 250 meters.

The sampling trap was a sterile plastic chamber containing four slides coated with beef broth agar, separated by 3 mm sterile balsa squares. The loaded trap was placed in a sampling capsule, Fig. 8-1. Ball valves in the nose cone and behind the trap allowed free air flow through the sampling chamber during the second stage of powered flight and the coast phase but closed and sealed the chamber during the recovery phase. In this way, exposure time was kept constant for all flights.

When constructing the capsule, I modified an Estes PNC-60AH nose cone by cutting off the tip to leave a 1.25 cm opening. I then added a ball valve support 4 cm below the opening; the support is two cross wires passing through the nose cone and epoxied in place. The upper and lower valves are marbles 1.9 cm in diameter. I used white glue for all cardboard-to-cardboard joints. The 1/16" and 1/8" aluminum tubes were epoxied in place; epoxy is the only suitable adhesive for these metal-to-wood and metal-to-cardboard joints.

The launch vehicle was a two-stage Estes Omega (No. 1200), using a D12-0 engine in the booster and either a D12-

5 or C5-3 (with adapter) engine in the upper stage. Two shock cords connected the capsule's nose cone cap to the booster; these cords opened the capsule after staging. Sampling began at about 50 feet, eliminating the possibility of ground-level contamination.

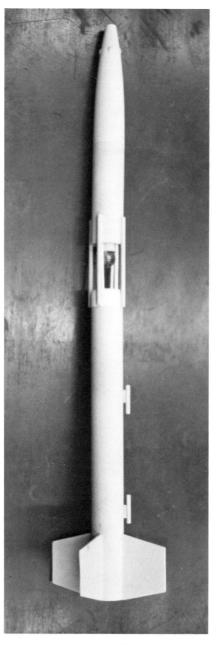

Identifying features of an atmospheric sampling rocket include a nose cone with an opening in its tip and another opening beneath the sampling chamber. Both openings were fitted with check valves, in this case ball valves made from glass marbles.

The capsule was recovered with a 76 cm-diameter nylon parachute. Tumble recovery was used for both the booster and launch vehicle.

Flight preparation and flight sequence

1) Laboratory preparation. Prior to loading the sampling chamber, slides were placed in Petri dishes and 1 mm of beef broth agar was poured over the slides. After the agar hardened, the slide pack with spacers was assembled, wrapped in aluminum foil, and placed in the plastic vial. Standard aseptic techniques were used throughout, and a control pack was prepared for each flight. The flight pack was then loaded into the capsule, the capsule capped, and the bottom vent plugged with sterile cotton. The capsule was then loaded on an Omega launch vehicle and the capsule shock cords connected to the booster.

2) Field preparations. On arrival at the launch site the rocket was placed onto the launch tower and the booster's igniter hooked up. A recovery crew of at least four individuals was deployed downwind. The capsule had to be caught by hand before it touched the ground to prevent possible contamination with soil bacteria. Just prior to launch, the sterile cotton plug in the vent tube was removed to permit free air flow. Wind speed, temperature, and barometric pressure were recorded prior to launch.

3) Flight profile. The flight profile can be broken down into five steps: a) Booster ignition and launch. b) Booster separation — sampling begins when staging occurs and the booster pulls off the nose cone cap. c) Sampling continues during the upper stage and coast phases. d) Firing of the ejection charge deploys the parachute; ball valves seal off the sampling chamber as the parachute opens. e) Following recovery, the nose cone and vent are plugged with sterile cotton.

4) Final laboratory procedures. In the laboratory the flight pack was removed from the capsule and each slide placed in a sterile Petri dish. The slides were incubated at 28 degrees centigrade for 48 hours and then examined for the presence of bacterial or fungal colonies.

Results and discussion

While no attempt was made to identify bacterial colonies growing on the sample slides, several types were noted

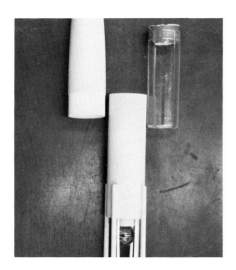

Nutrient agar-coated glass slides fit inside the clear plastic vial; the vial rested inside the rocket body tube between the upper and lower check valves. Standard aseptic techniques were used.

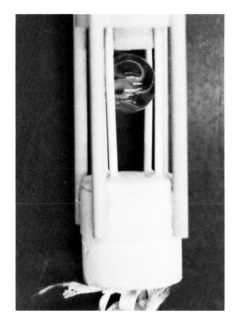

The lower check valve marble was held in place by an aluminum tubing cage. The chamber was usually lofted by a two-stage Estes Omega model rocket.

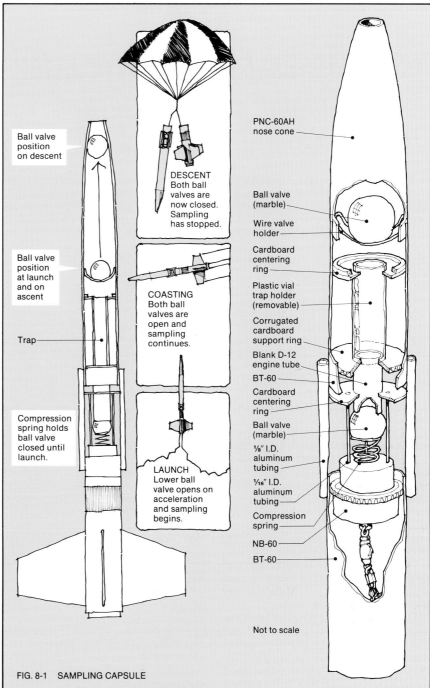

Ball valve position on descent

Ball valve position at launch and on ascent

Trap

Compression spring holds ball valve closed until launch.

DESCENT
Both ball valves are now closed. Sampling has stopped.

COASTING
Both ball valves are open and sampling continues.

LAUNCH
Lower ball valve opens on acceleration and sampling begins.

PNC-60AH nose cone

Ball valve (marble)

Wire valve holder

Cardboard centering ring

Plastic vial trap holder (removable)

Corrugated cardboard support ring

Blank D-12 engine tube

BT-60

Cardboard centering ring

Ball valve (marble)

⅛″ I.D. aluminum tubing

¹⁄₁₆″ I.D. aluminum tubing

Compression spring

NB-60

BT-60

Not to scale

FIG. 8-1 SAMPLING CAPSULE

during visual observation. The number of colonies per slide ranged from 0 to 23. There did not appear to be a correlation between environmental conditions and bacterial colony numbers in the 11 flights made with this sampling unit. Very few fungal colonies were seen.

Several pertinent observations can be made:
• Wind tunnel data is required to determine the actual air flow through the sampler.
• The Omega is an exceptionally reliable vehicle. There have been no failures in more than 30 flights.
• A checklist covering each step in preparing the sampler and launching the rocket is essential.

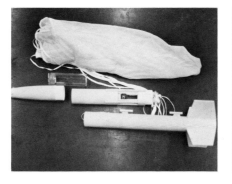

Here is a single-stage vehicle that also performed well. In both cases, a 76 cm-diameter nylon parachute ensured safe recovery of the sampling chamber, which had to be caught by hand before it touched the ground.

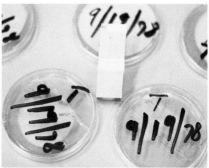

After each flight, the agar-coated slides were removed from the sampling chamber, transferred to sterile Petri dishes, incubated at 28 degrees centigrade for 48 hours, and then examined for the presence of bacterial or fungal colonies.

49

9. A low-cost impulse recorder for model rocket engines

By Henry A. Taitt and Charles E. Miller, Jr.

This homemade impulse recorder provides you with a method of measuring the thrust of a model rocket engine, a way to record small increments of time, and the means to physically tie them together while the rocket engine is firing — all at a price you can afford.

Assembly

Parts A through D can all be cut from a 16″ piece of 2 x 6 construction grade fir. Part E is ⅜″ plywood and F is a 2″-long ½″ dowel. The dimensions of parts A to E are:

A 1⅝″ x 4½″ x 4½″
B 1⅜″ x 4″ x 11¼″
C ⅝″ x 1⅝″ x 11¼″
D ⅝″ x 1⅝″ x 11¼″
E ⅜″ x 11½″ x 14″

(Part B has been planed down to 1⅜″ to match the height of the vibration timer.)

It's best to cut the notch into the side of D and B before sawing D from B. The notch is about 2″ from the end and is just wide enough (2⅛″) to accommodate the vibration timer. The cut goes to the center of B.

The oval drawn on E can be cut out to provide a handhold. The dowel, F, goes into a hole drilled in E and serves as a mount for the vibration timer — don't drill the hole until you've determined the timer's exact position.

Drill a ¾″ hole into the end of A. The hole is 3″ deep and centered ¾″ from one end of A. Drill a 3/16″ hole at the center of the ¾″ hole, passing all the way through the wood. (You may want to drill the 3/16″ hole first so that it will serve as a pilot for the ¾″ drill.)

Now prepare the hardware that will go inside the hole. Cut 6-32 threads for about ½″ on one end of a 6″-long, ⅛″-diameter brass or steel rod. At the end opposite the threads, bend the rod into a loop just large enough to accommodate a No. 6 machine screw. Insert the rod into A so that its threaded end protrudes from the ¾″ hole. The loop at its other end will prevent the rod from going all the way through. Slip a standard ¼″ (inner diameter) washer over the rod and let the washer slide into the hole — this should be a snug fit because the washer's outside diameter is ¾″. The washer serves as a metal plate for the spring to act against and protects the wood from flame damage caused by engine ejection charges.

The spring goes on next and represents the greatest challenge. It must slide freely in the hole, be less than .710″ in diameter, and compress at least 1½″ when subjected to a force equal to the maximum thrust of the engine to be tested. A spring that compresses about 1½″ when subjected to a force of 48 ounces (13.4 newtons) will work with most regular-size engines. Mini-engines require a compression of 1½″ when a force of 28 ounces (7.8 newtons) acts upon the spring. Most good hardware stores stock a wide selection of springs, so you should be able to find something suitable.

Screw a 6-32 nut onto the threaded end of the rod, followed by an SAE ¼″ washer (⅝″ outside diameter) and another 6-32 nut. These capture the spring between the two washers. Then insert a 4½″-long piece of Estes BT-20 body tube (inner diameter .710″, outer diameter .736″) into the ¾″ hole, sliding the tube over the spring and rod. The BT-20 tube is just the right size to accommodate a regular-size engine.

The holder for the marker pen (a Sanford Sharpie is shown) is made of thin sheet steel, brass, or aluminum; it's held to the rod by a 6-32 machine screw and nut. The screw passes through the loop you bent earlier.

Mount parts A, B, C, and D on E using screws and white glue or any other wood glue. Determine the timer's exact location, then drill a ½″ hole in E for dowel F, and glue F in place.

The vibration timer, called an Acceleration Timer (Catalog No. 16025), is available from Science Kit, Inc., 777 East Park Drive, Tonawanda, NY 14150.

The recording paper slides between C and D on the top of B; it is pulled by an HO scale model railroad locomotive. These train engines are inexpensive and their speed can be regulated by varying the voltage applied to the tracks through a model railroad power pack.

Use graph paper so that the pen's tracing will be easier to evaluate. Note the black strip along one side of the paper. This is a piece of typewriter ribbon taped to the graph paper. The vibration timer puts dots on the paper at the rate of 60 per second, thus giving a good time base with which to relate the engine's thrust.

Calibration

Calibrate the spring and graph paper by attaching weights of known mass to a string tied to the penholder end of the rod. The photo shows the string passing through fancy lab demonstration pulleys — your equipment need not be so elaborate. The mass in kilograms required to move the pen a known number of lines on the graph paper can then be multiplied by 9.8 newton/kilogram to find the force being applied. The force necessary to compress the spring (even if its compression is not linear), and thereby move the pen, can be found for each major line on the graph paper.

While calibrating, adjust the voltage to the locomotive to ensure that the dots are being spaced evenly and far enough apart that they can be easily related to the lines on the graph paper.

A piece of BT-20 body tube serves as a liner for the ¾″ hole in part A. The spring and the wire rod to which the pen is attached go inside this hole.

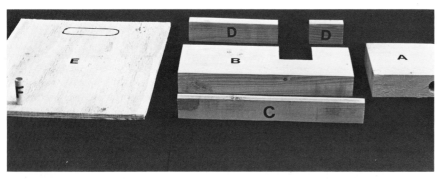

Most parts can be made from construction grade lumber or plywood.

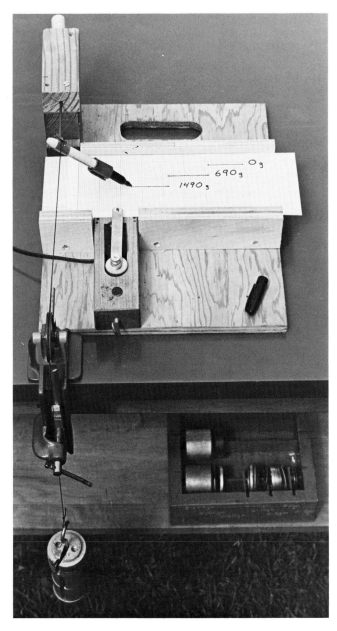

Before use, calibrate the recorder with weights of known mass, in this case 690 and 1,490 grams.

An HO locomotive pulls the graph paper, the vibration timer taps on the typewriter ribbon 60 times a second, the Estes B6-4 engine fires, and the marker pen traces the engine's thrust.

Clean the tracks for better electrical contact if the dots are not evenly spaced.

Engine tests and interpretation

Before firing, check that the rocket engine slides easily in the BT-20 tube and that the spring and rod do not bind. Be sure no one is standing near the engine during tests and keep in mind that the engine's ejection charge will kick the hot casing from the tube shortly after the main charge burns.

When testing, turn on the vibration timer first, then the locomotive, and ignite the engine about a half second later. The locomotive's speed should not be changing while the engine fires.

Note that the recorder is secured by metal rods stuck into the ground to prevent unwanted movement. All live tests must be conducted outdoors to prevent fumes from accumulating.

Here's how to interpret a typical engine test. The photo shows an engine's thrust trace on the longer sheet of graph paper. The shorter piece of graph paper contains calibration information derived earlier. The arrows and the letters A to H were added after the trace was made.

Look at the dots made by the vibration timer. The paper was at A when the locomotive started. By B the locomotive and paper had reached a constant velocity; there are 30 time intervals between B and C and another 30 between C and D. There are 58 lines on the graph between B and C and 59.2 between C and D. The closeness of the two numbers proves that the paper moved at a steady speed. Since 60 time intervals equal 1.0 second, and 117.2 lines passed in that time, the value T for each time line is T equals 1.0 second divided by 117.2 divisions or .0085 sec-

ond per division. The thrust calibration chart shows that adding 690 grams moved the pen 6.3 divisions, while adding 1,490 grams moved it 13 divisions.

The force per division can be found by: F = (1.49 kg) (9.8 N/kg) = 1.1 N/division ÷ 13 divisions.

Before the timer was turned on or the engine was fired, the train was run to produce the straight line from G to H (the leftmost dot at G indicates where the pen sat waiting).

When the paper was repositioned (the second dot at G) and run again with the timer marking and the pen recording the engine's thrust, it was reassuring to see line G to H retraced so closely in the regions G to E and F to H.

Since each square under the curve has a vertical side equal to a force of 1.1 newtons and a horizontal side equal to .0085 second, each square equals an impulse of 1.1 newtons times .0085 sec-

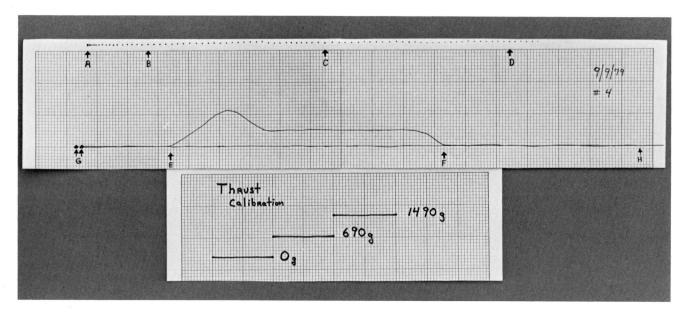

Use the thrust calibration chart prepared earlier when interpreting data gathered during an engine test.

TABLE 9-1
COMPARATIVE RESULTS WITH THE ESTES RECORDER AND THE LOW-COST RECORDER (ESTES B6-4 ENGINE)

ESTES RECORDER		LOW-COST RECORDER	
Peak thrust (Newtons)	Total impulse (Newton-seconds)	Peak thrust (Newtons)	Total impulse (Newton-seconds)
11.88	4.45	12.32*	4.7
10.62	4.39	11.50	4.6
10.62	4.41	10.70	4.3
10.31	4.17	10.40	4.1
9.38	4.17	9.60	3.9
Average			
10.56	4.32	10.90	4.3
Standard deviation			
.90	.14	1.05	.3

*This engine's trace is shown in the photo above.

ond or .00935 newton-second per square.

The total impulse is the number of squares under the curve and above E to F (call it NUM, for number) times the impulse per square, so that total impulse equals .00935 newton-second times NUM.

The impulse for each time interval can be found by counting the number of vertical squares for each time interval and multiplying that by the impulse per square.

Table 9-1 shows the results of ten test firings of Estes B6-4 engines, five with the low-cost recorder, five with much more sophisticated equipment used at Estes.

As you can see, the results are close enough to make our inexpensive recorder very appealing.